Power Quest

Power Quest
The Journey into Manhood
Carol Batdorf
Illustrated by Tracy Williams Cheney
hancock
house

ISBN 0-88839-240-0

Cataloging in Publication Data

Batdorf, Carol, 1919-
Power Quest
: The Journey into Manhood

Sequel to: Spirit Quest.
ISBN 0-88839-240-0

1. Cha it zit, Chief of the Lummi - Fiction.
2. Lummi Indians - Fiction. I. Title.

PS3552.A72P6 1990 813'.54 C90-091108-5

I dedicate this book to Cynthia Wilson of Lummi, who, as a friend, inspired me to write and to continue to write about the Salish people of long ago. C. B.

For my grandma — whose love of books shaped my own. T. W. C.

Edited by Cheryl Smiley
Production by Cheryl Smiley and Phil Della

Printed in Hong Kong

Published simultaneously in Canada and the United States by

HANCOCK HOUSE PUBLISHERS LTD.
19313 Zero Avenue, Surrey, B.C. V3S 5J9

HANCOCK HOUSE PUBLISHERS
1431 Harrison Avenue, P.O. Box X-1, Blaine, WA 98230

Contents

List of Maps

Prologue

We, who have colonized this great land we call America, have gloried in our achievements. We pride ourselves for turning the wilderness and open spaces into what we call productive lands. We marvel at our cities, our freeways, our universities, our industry. To have done it all in 200 years is indeed an achievement, but somehow we seem to have overlooked the fact that this great space wasn't empty when we came. For centuries, it supported a vital culture, one that produced languages, customs, moral codes, and challenges for its original people. There was much of value then.

Leaders arose then, as they do now, to preside over the changes and great events, and to leave a legacy for others who follow. The Lummi Indian leader known as Cha it zit, was such a one. He was born into that Salish group, known as the Nuh Lummi, before white men were familiar to them, and he lived through the culture clash to sign the treaty committing his people to reservation life. His story intrigued me greatly. I wondered what his boyhood was like, how manhood sat upon him, what contact with Europeans did to him and his people.

Piecing together parts of stories told to me by Lummi elders, studying the writings of explorers and early pioneers, and using the knowledge gained through personal contacts, plus employing a very lively imagination, I began to construct a picture in my mind of what life could have been like in the Pacific Northwest in the early eighteen hundreds when Cha it zit was born and lived.

My first book, *Spirit Quest*, reflects those early years when Cha it zit undertook the spirit quest (a sort of initiation), in which he was successful and was rewarded with the powerful wealth-giving spirit Tiolbax as his helper and protector. As Cha it zit underwent his time of testing, to determine if his spirit power was valid, his great-uncle and mentor Tselique – a great shaman – disappeared. This second book brings Cha it zit into manhood starting with him and his grandfather returning from a search for Tselique.

Some of the characters are real, Cha it zit, his father Chief Sax ump ki, and his brother Cha wentz. Their stories have been handed down. Other characters are imaginary, but all are based on the way such people would have behaved and lived in those days in that setting.

I have great admiration and respect for the old Indian ways, so far as I understand them. If I have been open and frank about some aspects of their lives it is only to create a complete, believable account. I have endeavored to delineate life as it was lived, with its challenges and rewards. Not all of it was beautiful, but neither has any period of history been so, nor even our own lives today. I have also tried to stress what was important to those people of long ago, what motivated them to act and think as they did.

Power Quest is essentially a historic novel, but one based on real characters and real events. As such, it is my earnest hope that it will be enjoyable to read. If it informs, too, so that readers will have a look at the life that existed here before "we came" and "conquered" the land, I will have fulfilled my goal.

CAROL BATDORF
Bellingham, Washington

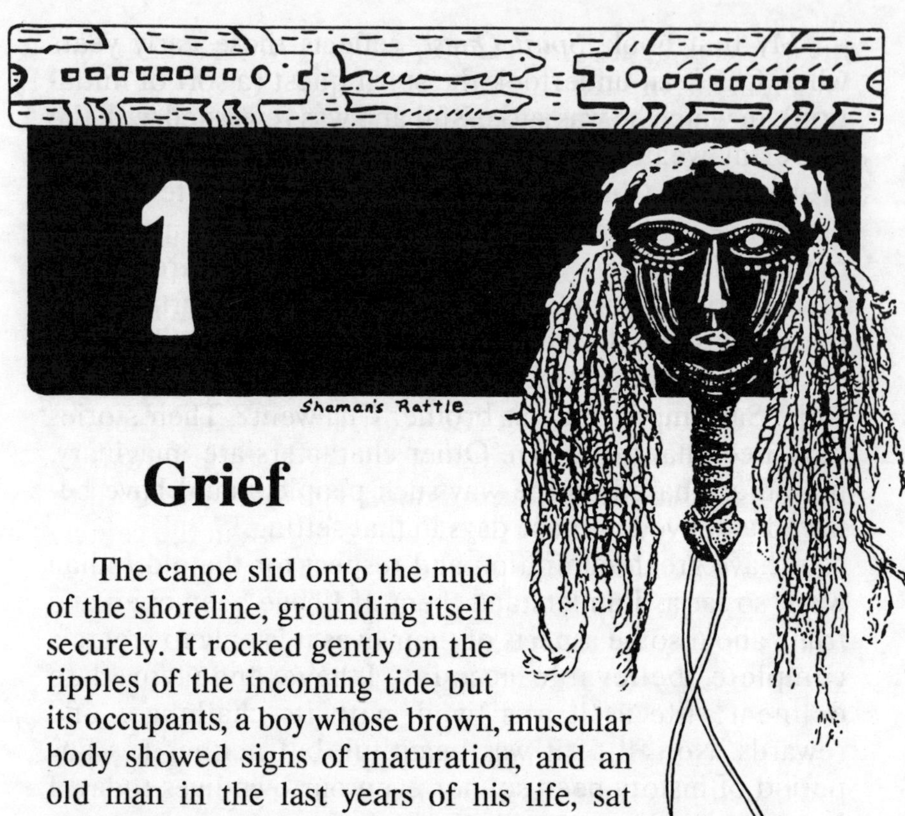

Grief

The canoe slid onto the mud of the shoreline, grounding itself securely. It rocked gently on the ripples of the incoming tide but its occupants, a boy whose brown, muscular body showed signs of maturation, and an old man in the last years of his life, sat quietly.

Neither spoke. The man, with red-rimmed eyes, stared at a distant snow capped mountain which rose above a dense forest of cedar trees backed by ever-receding hills. The boy seemed to see nothing, his eyes riveted on the ever-moving water lapping at the canoe prow.

Beyond, on a rise above the beach, the village called Swetquem seemed to slumber. Smoke curled loosely, lazily, from the smoke holes of a line of houses, but there was no sound except the distant call of a raven and the lapping of the sea. Time passed. The sun, which was hanging low on the horizon over the sea, turned the mountain to gold and the hills faded to mauve, but the two did not stir. Then, as the sun sank behind the islands, the air turned cold, bringing them out of their reverie.

"It is time to tell them, grandfather," the boy-man said.

"Oh—yes," the old man agreed with a sigh. "It is a hard thing to do. Help me, Cha it zit. My bones are stiff and cold."

Leaping from the canoe, the one called Cha it zit half carried his grandfather across the sand and beach rocks to the path that led up to the village.

"Get me a stick. I'll walk from here," the old one ordered. "No one will see Sta nek carried by his grandson."

They entered the largest of the cedar plank houses, stooping to pass under the low door. The interior was quite dark, except for the fires which glowed down in the center. Dimly they could see the people assembled about the fires, women preparing the evening meal, children playing, men talking as they worked on their fishing lures.

Then a voice called, "They're back. Cha it zit and Sta nek are here."

A powerfully built man rose from among the gathering and called "Tselique? What of him?"

"You must tell them," the boy said softly to Sta nek. "It is better that you, Tselique's brother, tell."

"Yes . . ." The old man seemed to be utterly bemused. "Yes . . ."

By that time the people had gathered around the pair. Their eyes questioned, but no one spoke.

At last an old woman took Sta nek's hand. "Come," she said firmly. "Come by the fire. Warm yourself, husband. Then, when you are ready, tell us of Tselique."

Mechanically, Sta nek allowed himself to be led. Cha it zit followed.

When the old man had been seated on a bentwood box near the farthest fire, he seemed to become more aware; to come back from a distant place.

"Tseutsi, my wife," he spoke. Then he shook his head and unabashedly he wept. No more words would come. His old frame shook, convulsed by his sobs.

He need say nothing more. The people knew why he cried. It was a shocking thing to see him, Sta nek, their elder, once the strength of the family, weak as a child and sobbing out in grief.

"He is dead," Tseutsi shrieked, "Tselique is dead." She reached into the fire pit, drew out a handful of ashes, and smeared her face and body. "He is dead – gone from us!" Her cries were shrill and wild as the north wind. Other women joined their voices to hers. The children ran frightened to huddle together in the dark outer reaches of the great house.

Then men, equally affected by the tragic news, swayed and moaned in their places.

Cha it zit, deep in his own private grief – for Tselique had meant much to him – hunched close to Sta nek, who had been the dead man's brother and close friend. Cha it zit did not attempt to speak. No one would have heard him above the mourners' wails, and his feelings were too deep to share.

He remembered it had been Tselique who guided him through his spirit quest, who intervened with his own *spirit power* when Cha it zit needed help, who gave him a powerful amulet and shared his shamanist knowledge to help him succeed. The debt he owed the old shaman, his great-uncle, was beyond measure.

"I will repay him," Cha it zit swore to himself. "By my life and my deeds, I will repay him. I don't know how but I will."

Then the boy felt a hand on his shoulder and looked up into his father's face. Sax ump ki. The very name inspired respect among his people, and of them all, Cha it zit admired him the most.

"My son," Sax ump ki addressed him, "I know what you are feeling. He was special to you as you were to him. It is no disgrace for a man to wail. It releases the soul and makes it easier to bear grief. I would know what happened to my uncle, but not yet. Cleanse yourself first. Both you and Sta nek must remove from your bodies any scent of him so that his soul will not be attracted to you and stay here rather than journeying to the land of the dead. Harm could come to you and us – for his powers were great."

"Father," Cha it zit choked out the word in a strangled voice. "I cannot cry, my throat is tied, my being is tied."

"Come," Sax ump ki pulled the boy to his feet and forced him out of the longhouse. Sax ump ki pushed Cha it zit toward the salt water, stopping to break off a cedar branch along the way.

"Now," he ordered, handing him the bough, "cleanse yourself."

As in a trance, Cha it zit obeyed. The cold water penetrated his consciousness. Viciously, he beat himself until his body glowed red from the icy water and the harsh bough.

Then his mind responded. "Tselique," he cried out in a strangled voice.

"Do not use that name again," Sax ump ki admonished him. "Let his soul go to the place that is waiting for him. Let him go."

"Oooooh," the boy screamed in reply. The sound of his grief vocalized released him. He cried out again and again. He felt better – relieved.

"Come now," Sax ump ki said kindly. "It is done."

News of the death of Tselique, the old shaman, spread quickly through the village. By the time Cha it zit and Sax ump ki returned to their longhouse, it was filled with wailing, mourning clansmen.

All the women had smeared their faces with ashes and were raising an awesome noise. They contorted their bodies, twisting and writhing, working themselves into a frenzy of emotion. Droning beneath their shrieks, the men intoned a bass to the crescendos.

On and on the lament continued, until exhaustion forced the people to rest. Gradually, except for a few sporadic gusts, the room became quiet. That was the moment that Sax ump ki had waited for. He mounted the platform at the end of the room, his accustomed place, and rapped on the floorboards with his talking stick.

The gathered clan gave attention, waiting to hear what their leader would say.

"This is a sad time for us," he began. "We respected my uncle. He was ready to serve us, to protect us from harm. His power was great. Who among us has not profited from his efforts?"

Sax ump ki looked directly at each of his people and waited for his words to be felt. No one moved. Then he continued.

"We do not know what my father, Sta nek, and my son, Cha it zit, found or how they cared for his body. If he is able, I ask Sta nek to tell us."

The old man raised his head and looked at Sax ump ki with tear-reddened eyes. His hands trembled.

"I – I cannot speak of it," he said. "I ask Cha it zit to speak for me."

Sax ump ki looked at his son. As a frog leaps, Cha it zit felt his heart leap into his throat. Then he rose and mounted the platform. Strength came from deep within himself – the power to speak about the unspeakable. He took the talking stick from his father's hand, and then wondered how he found the audacity to do it. Understanding, Sax ump ki let it go.

"This is how it happened," he began. "When Sta nek and I reached the island, we went to the old longhouse that my great-uncle stayed in. We went inside and found him lying under his blanket, dead. At first we dared not touch him, but then we saw that he had been dead for a long time and his soul had probably already gone away. We washed him and wrapped him in his cattail mats and put his body in a tree, as is proper. There were some things in the house that he had given to me. I hid those so no harm would come to them and they could do no harm to others. Then we came home."

The women, hearing the details, began to wail and moan anew. Sax ump ki took back his talking stick and rapped loudly. The noise ceased.

"We will leave my uncle where he is, in the place he chose to be. If we do not entice it, his spirit will stay there. It can harm no one and, when he is ready, he will go to the place of the dead. Would anyone here speak?"

The group stirred restlessly; people looked at each other. At last a woman's voice was heard, old and quavering. "Tseutsi, my mother," Sax ump ki said.

"He was our friend," Tseutsi began in a small voice. "My husband cannot talk of it, his voice is not in him, so I will speak for him. Although my husband's brother is not with us, we will remember him. This was his." She held up a wooden box, carved

Soul catcher and spirit board

in a simple design with the patina of age and use upon it. "I will send it to him."

The old woman rose and placed the box on her fire. Immediately the dry cedar caught and flames leaped into the air.

Cha it zit seemed to see a hand in the flames take the box as it turned to glowing ashes.

Others rose and put articles in the fire: a token Tselique had given away, a mat he had used, a ladle, a fur cape.

"It is time we ate," Sax ump ki announced when the burning was done.

The women wiped the ashes from themselves and busied themselves around the fire. Food was for the living, and it must be prepared before it could be eaten.

Food was for the dead, too. Each who partook of the meal put a morsel into the fire for Tselique to sustain him until he had found his way along the path to the netherworld.

The mourning continued, but it was sporadic, fused with the everyday tasks that must be performed if the people were to survive.

Cha it zit found it hard to let Tselique go, as he was told he must. He could not exorcise his grief with wailing, as the women did. His brief outburst at the clearing was his only outlet. So he lived with it.

When his work was done – which was seldom – he found respite on a rocky ledge overlooking the bay. There, watching the sea gulls scavenging for food, or the harbor seals at play, or his own people attending to their chores, he found peace. Often, a raven which had formed an attachment to him hovered nearby, occasionally calling in a mellow, "Aawk! Aawk! Aawk!"

There the boy thought of the spirit quest that Tselique had guided him through, of the terrible dives he had made to find his spirit, *Tiolbax*, of his testing in the smokehouse

when his power was first felt. Tselique's power was there, too, he knew. When did the old shaman die? Was it when he appeared in the flames? Was it too much of a drain on his old body? All that *squidelich,* did it kill him? "Oh no – deny that thought," his mind rebelled.

Cha it zit was sitting on the ledge, a few days after the burning, when Sax ump ki came to him. He was surprised but also embarrassed to be honored by such a visit.

"I was only resting for a minute," he started to say, rising.

"I wish to talk to you," Sax ump ki motioned for his son to stay. "There are certain things we must discuss."

Cha it zit sank back to the rock and waited.

"You are becoming a man. Soon you will be called upon to fill a man's place among our people. It is time that we consider a suitable wife for you."

"Wife?" Cha it zit asked in amazement.

"You are young, yes. But it takes much effort to arrange a marriage. Sometimes many moons pass before all the proper rituals can take place. Already your older brother Cha wentz has a wife. Now we must seek one for you."

"There is one I would consider," Cha it zit answered, speaking very deliberately, his face reddening.

"Oh?" his father asked, with a peculiar inflection in his voice, one that Cha it zit did not like to hear. "You would consider?"

"I only thought – "

"It is not for you to decide," his father interrupted. "Your mate will provide an alliance for us with another village, another clan. She must possess wealth and power and prestige through her lineage so that we will have access to it."

"Yes."

"So," his father said, his voice tinged with sarcasm, "who is this maid that you would consider?"

"She is the daughter of the headman of the Samish people."

"Ah, yes," Sax ump ki answered thoughtfully. "Yes."

Then, abruptly as he came, he rose and left.

Cha it zit hurled a rock into the water. Then he threw another, and another, with such vehemence that the sea gulls scattered and the seals dove beneath the waves.

"Married!" he muttered. Then he thought of the girl. He had only seen her once, at his naming. But there was something about the way she carried herself, something about the depth of her black-brown eyes when she had looked at him and said "Cha it zit" that he had always remembered. "Maybe it wouldn't be so bad, after all," he thought.

He was about to leave when he spotted something out in the water moving rapidly toward the shore.

2

Yakultas

Cha it zit lunged into action.

"Yakultas!" he cried, racing toward the village. His shouts set the dogs that skulked about the houses to barking. "Yakultas coming!" he shouted.

He met his brother, Cha wentz, who had gained a reputation as a warrior, running toward the beach where canoes lay in readiness. Others, stalwart, muscled men, proud of their prowess, were emerging from houses, grasping spears, bows, and arrows as they ran.

Out in the straits the canoes, now six of them, grew steadily larger.

Then Sax ump ki was at the boy's side.

"You are not a warrior," he said. "Go with the women and children to the forest. Stay with them; protect them with your life, if need be. Don't let them be captured." Then he muttered, "Better they be dead than be slaves."

"Here," he thrust a spear in the boy's hand and hurried on down to the beach.

Already the warriors were shoving their canoes out into the bay. The dreaded enemy from the North must not reach the village. An open battle at sea, with the risk of drowning or perishing from spears and arrows at close range, was preferable to facing the Yakultas on land where they could do great harm by stealing supplies and taking women and children for slaves.

Chanting their power songs, the men lunged at the paddles, churning toward the approaching enemy.

Cha it zit watched for as long as he dared, and then, spear in hand, raced toward the houses, calling out, "Hurry, run to the woods!"

Latsi, his sister, was first out of the low door. She carried a basket and held a small girl by the hand. Then came his mother, Wanana, with an armload of blankets and a back basket suspended by a trump line from her forehead. Tseutsi, hobbled out carrying mats and urging several children before her. Looking in, Cha it zit saw that there were others, wives of Sax ump ki, putting together bundles of necessaries. Amid the confusion sat old Sta nek.

"Hurry," Cha it zit called. "Come grandfather, Sax ump ki wants us to go into the woods and hide."

"Let them go," Sta nek said proudly. "I cannot hide like a woman. My spear will find a target or two if the Yakultas come here."

Cha it zit knew it was useless to argue so he urged the others to hurry and raced out to lead the way. He knew just the place to take the vulnerable people.

There was a small gully where a stream flowed through. Branches and bushes hung in a dense thicket over the water, making an excellent place to hide. He had often gone there to play with other boys when he was younger.

He found the women and children assembled behind the houses, waiting for him and the stragglers.

Cha it zit grasped the spear—he felt strangely exhilarated. It was a good thing to feel important. When all were present, he led them on a faint trail, one he had once used but that was now overgrown by the jungle-like growth natural to the area along the Pacific Northwest inland waters.

He was careful not to disturb the foliage, and he warned the others to step lightly. No sound was made as they eased along. Even the littlest child felt the danger that could follow their tracks if their warriors were defeated.

When they reached the place Cha it zit had in mind, he parted the dense underbrush carefully and eased through, holding the branches and barbs back for the others. Then, while they waited, he tramped and pulled bushes till his hands and feet were bloodied by the thorns and stubble. In the small clearing he made, the group settled down on the mats they had brought. They wrapped blankets around themselves to ward off the chill and they waited.

If a baby began to cry, its mother pinched its nose and the mewling stopped. Baskets were opened and dried berry cakes were passed around. All this, but no sound.

Cha it zit thought of the sea battle that was probably taking place in the waters near the village. The pattern had been set long ago. As the canoes neared each other to exchange spears and arrows, the canoes were tilted so that a broad surface was presented to the enemy, forming a bulwark of wood which the warriors crouched behind. Occasionally, an aggressive crew would attempt to ram an enemy canoe and tip the warriors into the icy waters. Then they were easy to spear. It was the mark of a great warrior if he took home a scalp or an enemy head, so hand-to-hand combat, while difficult in dugout canoes, was not out of the question.

Cha it zit knew, too, that such battles weren't a matter of complete annihilation. The attackers would often turn and run if the battle was going against them. But they could be expected to return another day.

"Those Yakultas have to be dealt with," Cha it zit thought as he sat with the women and children. He recalled tales of how they and other fierce Northern raiders had swept into the channels his people used and killed so many warriors and stolen so many women and children that whole villages had been deserted. The old men spoke of as many people as there were stars in the sky in the old days.

"Now there are only a handful of what once was here," they would say.

He remembered the story that his great-uncle, whose name he could no longer mention, had told him. That was when Sax ump ki was a young man, long before Cha it zit was born. Northerns had swept onto an island where his people had assembled to dig clams and dry them. The men were away on a hunting and fishing expedition. The Yakultas had killed the old people and taken the women and children as slaves.

When Sax ump ki learned of the disaster, he swore vengeance. He had his revenge when the Yakultas came again. Lummi warriors had seen them coming and had made a trap for the northern marauders. Sax ump ki had allowed one Yakulta to live to return to his people to tell them of the might of the Lummis. But even then, the raids did not stop.

Cha it zit knew that his father worried about the violence brought by the Yakultas. He recalled how his father would pace the beach, looking out to the sea. "They must be stopped," he would say. "I must think of some way to keep them away from our villages."

"So far, he has done nothing," Cha it zit thought. "I wonder what he can do." His mind dwelled on possibilities, but he found no acceptable plan. "I'm not a warrior," he conceded. "It is not my concern."

All was quiet. A small bird hopped on a branch nearby and the boy could hear scurrying noises in the thicket. Then a raven cawed overhead. Wherever he went, Cha it zit noted, the raven seemed to be. He accepted the bird as a symbol of his *spirit gift* and, somehow, as a legacy from Tselique. The bird's presence comforted him. "So, I am not a warrior," he told himself, "I still have a great gift. Some day, I'll know what it can do for me."

Time passed. It seemed an eternity to the group who cowered in the bushes. They dared not sleep, so they sat in the growing darkness, simply waiting.

Suddenly, there was a noise – more a rustle. Every sense Cha it zit had sprang into a state of readiness. He sat still, his body tense, and he clutched his spear.

"Cha it zit," a voice said, "are you there? It's over, come out."

Cha wentz pushed through the opening. Cha it zit relaxed at the sight of his brother's face, but he still held the spear tightly. He saw sarcasm twist Cha wentz's features.

"So," the warrior sneered, "you would hide with the women? What would you do with that weapon, brother, if you were faced with Yakultas instead of me? Slay them all? Ha! I'll wager not one would feel the sting of that barb."

He hit the speartip with his fist, knocking it from Cha it zit's grip.

"If you had been the enemy your belly would know what my spear could do," the boy spat out the words angrily.

"Ah! The small one has a temper," Cha wentz laughed. "Come now," he beckoned to the others, "it is safe."

As they were returning through dense thickets of blackberries, nettles, giant sword ferns, wild rose bushes, and vine maples, Cha wentz spoke to Cha it zit.

"We are not alike," he began. "I do not understand your ways."

"Perhaps because we have the same father, but different mothers," Cha it zit ventured. "I do not understand you, either. How can you wish to kill?"

"Wish!" The word exploded from Cha wentz's mouth.

Cha it zit cringed involuntarily.

"Wish," Cha wentz repeated. "It is my obligation to kill. My spirit is that of a warrior. It is what I live to do." Then he looked directly at Cha it zit and growled, "Where would those like you, brother, be today if there were not great warriors like me to meet the enemy. I'll tell you. You would be a slave in a Northern village. A Yakulta slave, cowering and crawling before a hard master, or your head would be decaying on a pole, for all to see what became of the son of a Lummi leader."

Cha it zit, while the words were distasteful to him, thought it best not to answer. They walked in silence, threading their way along the faint trail with the women and children following.

"Consider this," Cha wentz said. "It was very unwise to hide together, as you did. If the enemy found you, he would find all. Scatter, so some may live."

"How did you find us?"

"It was not hard. I know of your hiding place. I used to follow you sometimes when you were little."

"You spied on me?" Cha it zit asked, amazed that his brother took that much interest.

"I wanted you to be safe," he said gruffly. Then he continued, "Also, it is impossible for many feet to pass along a way and not leave the ground trampled. Anyone could have found you."

"What happened out there?" Cha it zit asked, changing the direction of the talk, which was becoming more unpleasant to him as it continued.

"We paddled furiously, shouting our power songs so that they echoed across the water. There were fifteen canoes filled with our warriors. We made quite a sight, I'm sure. When we neared the enemy, they saw that we were fierce and filled with our spirit power. They turned and fled. We chased them for a while to let them know we were ready for battle. It was a victory for us."

"Will they come back?" Cha it zit asked.

"Perhaps," Cha wentz answered.

In the longhouses life resumed its normal pace. Hearth fires were fanned. The same women who had recently huddled in fear calmly resumed their household chores.

Cha it zit noted that Tseutsi had cut her hair. Using a sharpened mussel shell, she had sheared her graying braids and now her hair hung to her shoulders.

"It is a mourning custom among my people," she explained to him. "Tomorrow I will cut it to below my ears. All will know that I mourn for my husband's brother." When she had finished speaking, she sat on her mat and began to moan. Soon other women joined her and once again the longhouse was filled with the sounds of sorrow.

After the evening meal, the headmen of the village, one to each house, five in all, gathered at Sax ump ki's place on the platform at the head of his house. Soon they were joined by Cha wentz and other warriors. Cha it zit was not invited to join the group. For some time he watched from a distance, wondering what they were discussing.

Latsi, his sister, several years younger, approached Cha it zit "Brother," she began humbly, "I would talk to you."

"The elders say it is not seemly for me, almost a man, with spirit power – even if I am not sure what it will do for me – to spend time with a girl," he answered miserably, for he had a strong affection for her.

"I know," she said. "But I want you to know that I thought you were very brave today. Don't worry about what that Cha wentz says. What do you think they are talking about?" she asked, motioning toward the group on the platform.

"I wish I knew," he answered.

"Why don't you go up there?"

"Me?" He was amazed she suggested such a thing.

"Maybe they would let you stay. Remember, you are Sax ump ki's second son."

"I don't know—" he began and then stopped. "Maybe I could get close enough to hear."

Latsi moved off toward the fire where Wanana was clearing away the remains of their meal and tossing scraps to the waiting dogs.

Slowly, Cha it zit inched toward the head of the house. He took advantage of the shadows which enveloped the sleeping platforms that ranged down both sides. The voices were becoming almost intelligible. He could recognize the speakers, but not what they were saying.

A hand grasped the nape of his neck and drew him erect.

"See what I found," he heard Cha wentz say. "This one wants to know what we are planning."

Cha wentz forced Cha it zit into the light and held him for all to see.

"Let him go," Sax ump ki commanded. Then he addressed the others. "My other son wishes to be included. He is young, and he is not of warrior blood, but he is wise, and, as you know, he found a powerful spirit helper. Do you permit him to remain?"

The headmen, wise, powerful elders who looked to Sax ump ki for leadership, sat without speaking, considering in their minds the wisdom of including such a one as Cha it zit in their plans.

Then one spoke, "Let him stay. These warriors of Lummi are an impetuous lot. Perhaps another voice should be heard."

"Yes," each agreed in turn.

"Sit with us, my son," Sax ump ki invited.

Cha wentz glowered when Cha it zit took a place among the leaders.

"Now," Sax ump ki said, "let us continue. Ever since the terrible massacre, when the Yakultas attacked our women and children while we warriors were away hunting and fishing, I have been plotting in my mind a way to protect our people. There have been too many attacks since that time. Even since the great battle when we defeated them and let only one warrior survive to tell of our power, even since that time, they continue to come."

"Yes, yes," the leaders murmured. "Something must be done. But what?"

"I have a plan. I have been waiting until my sons were ready to help. You can see that Cha wentz has a powerful warrior spirit. He can be depended on to fight to the death, if need be. I believe that Cha it zit also has strong power gifts. He is ready, too. You also have sons who are eager to defeat our enemy."

"We would hear your plan," one of the headmen said.

Sax ump ki nodded. "We will build a strong defense at Temxwigqsan, along the straits through which the Yakultas come. Inside, we will be safe, but from it we can launch a counter attack. We will place lookouts to warn us when enemy canoes are spotted. Then we will send runners to all the villages so that their warriors can rush to our defense. Our canoes will be hidden so that the Yakultas will believe that our defenders are away. We will always be ready."

Sax ump ki studied the faces of his listeners as if trying to see through their expressionless faces and into their minds. The elders seemed neither to accept nor reject his proposal.

"Do you recall the way we defeated the Yakultas before? How we lured them onto the beach and slaughtered them by closing in behind them. We warned them then, but they did not learn." Sax ump ki's voice rose to a fevered pitch, his eyes widened with hatred. He slammed his fist onto the hard-packed floor. "We will do it again!" Sax ump ki's words echoed through the longhouse, but the seated elders remained impassive.

"It will mean much work," one elder commented.

"We will have to move our village," another observed.

"Will all the people agree to the plan?" a third questioned.

The group expressed serious concerns. The men pondered and considered. At last, reflecting impatience, Cha wentz raised his hand. All regarded him with the respect due a warrior.

"You know my father," he began. "He has earned your admiration. When he speaks, men listen. Why do you doubt his wisdom now? Don't you realize that the Yakultas will return? They will bring many with them, enough to eliminate our villages and our people. Unless we stop them now, we will have no peace. We will be forever fearing for our lives and for our families. I say let's follow Sax ump ki's lead. Let us build the fort he asks for."

Suddenly Cha it zit admired his brother. He saw him anew, as a person changed from the one he had always known. It was a strange feeling. There stood Cha wentz, magnificent with his broad, bronzed shoulders and powerful warrior's chest, his arms muscled from paddling the big war canoes, addressing clan leaders, reasoning with them. Yet this was an unsettling experience, one the boy found hard to accept. But it was true, and Cha wentz became esteemed in the eyes of Cha it zit. "My brother," he thought, "I almost wish to be like you." Then he quickly reconsidered. "No, I have my own gift, my own way."

One of the participants nodded. "I have heard of the plight of the Semiahmoo people to the north of us. Once they were numerous. Their villages consisted of great houses. Then the Northerns, the Yakultas, came and killed or captured many of them. Time after time they came until now only a few live there. Many who survived have gone to another place. We do not want that to happen to us. I say, in the spring let us build our defense."

"And the rest of you?" Sax ump ki asked.

"I will speak to my kin-group," one said.

"If it was just my decision, I would say 'yes,' " another replied, "but I have many sons and nephews who will become involved. I will consult with them."

"As we know, Sax ump ki is a wise and respected man. What he proposes, I and my family will agree to," yet another stated.

Sax ump ki smiled at the last remark and then spoke. "You have heard the plan. Take it to your house, think on it, consider it, discuss it. In three days, return with your own plan or agree to mine. We must act if we are to survive."

So the matter was concluded for the moment. The *siems* rose, wrapped their blankets about themselves, and departed into the night.

The wailing began again in earnest. Voices rose to a screech and then fell to a moan. Bodies rocked. Hours later, gradually, the din subsided as mourners sought their pallets and fell asleep.

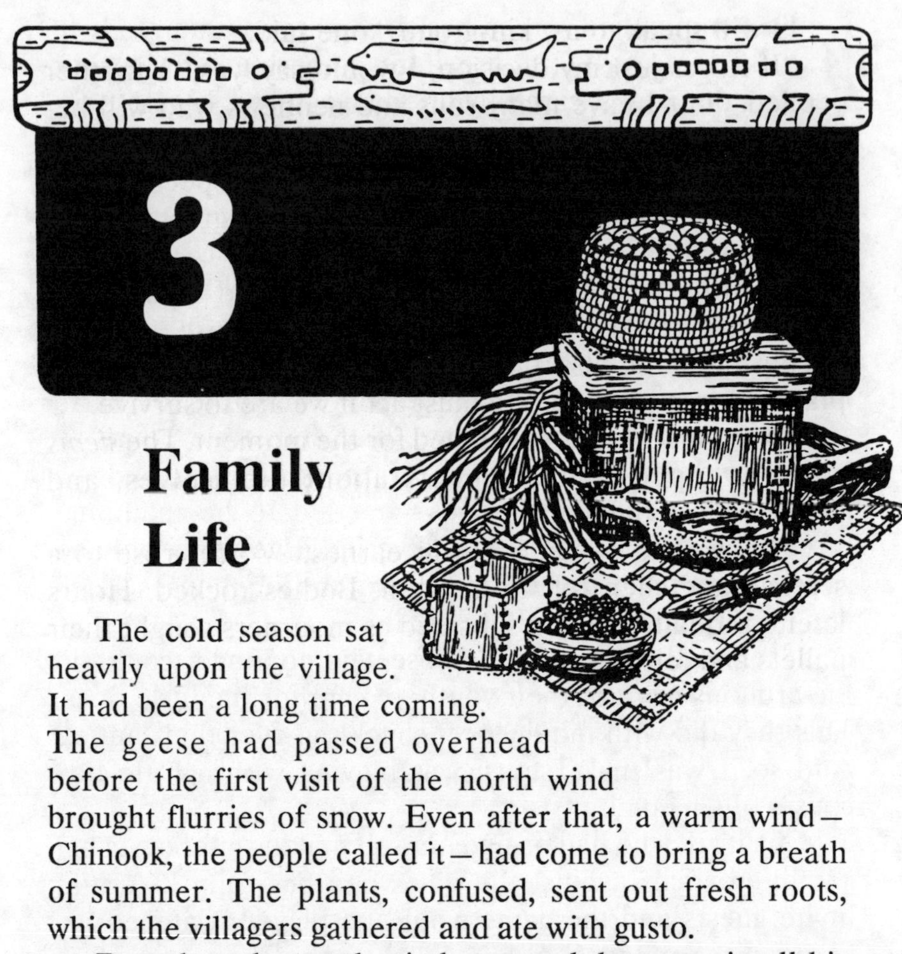

3

Family Life

The cold season sat heavily upon the village. It had been a long time coming. The geese had passed overhead before the first visit of the north wind brought flurries of snow. Even after that, a warm wind – Chinook, the people called it – had come to bring a breath of summer. The plants, confused, sent out fresh roots, which the villagers gathered and ate with gusto.

But when the north wind returned, he came in all his savage power. He shrieked around the wooden buildings tearing loose the heavy cedar roofing and puffing his way through cracks between the boards, billowing the cattail mats that the women had hung as insulation.

The people huddled around their fires inside, heaping wood on the flames to warm the area around the hearth. Even though they were housebound by the antics of the north wind and snow was piled outside the door, everyone was busy. There was always something that needed doing, it seemed.

The women had spread basketry material out, ready to be worked into baskets for the food gathering in the spring. Bundles of cattails, carefully dried and stored, lay close by. Old, worn mats needed to be replaced so that there would be shelter for those who camped on the islands while gathering camas and clams and for those who went up in the mountains for blueberries, mountain goat, deer, and elk.

The fishermen laid out their nets for mending with nettle twine, or worked on fish hooks, different kinds for the lordly salmon, mighty halibut, or reef fish. Hunters made new bows, laminating rawhide to yew wood for extra strength and resilience. Arrows, too, and stone points to adhere to them, must be made. Harpoons for seal and sea lions – with detachable heads – needed to be manufactured. The tasks were many and varied. Everyone, old and young, had a place in the family life of the longhouses. Each was important to the welfare of all.

It fell to the old women, those who no longer could do the arduous work of their youth, to care for the little ones. This they did with indulgence, chuckling over their antics. And so it was that Tseutsi was playing with a little boy cousin of Cha it zit.

Cha it zit, who had busied himself making a set of sla hal gambling sticks against the day when Sax ump ki would invite guests and the game would be played, watched his grandmother and noted her delight when the tot touched her face or smiled at her.

"Kake," she would call him. All baby boys were Kake till they received their first baby name.

"Don't you think it is time to give him a pet name?" he asked, not addressing Tseutsi directly, which would have been improper.

"When he does something special, or something unusual happens," she answered, speaking to no one in particular, "then Kake will have a name."

"So you will wait? Suppose nothing happens? Will he always be Kake?"

Tseutsi grinned, wrinkling further her lined face. "He must find a name soon. There will be another baby and we can't have two Kakes. Who will know which one we mean?"

Cha it zit looked about the longhouse. Who did Tseutsi mean? There was his mother, Wanana, working on a basket, looking healthy and contented, and his father's other wives, each preoccupied at her own work. There were several wives of Sax ump ki's brother, and the slave women, all engaged in the difficult work of splitting soaked cedar limbs at the far, drafty end of the house. Any one of them could be expecting, but Tseutsi would hardly be expected to care for their children. Cha wentz had married; could it be his wife? She was a pitiful, shy woman, hardly more than a girl, frightened and homesick for her people. No, it couldn't be her. When women were to have babies, he had observed, they usually became robust and smiled a lot.

"Would it be that Wanana swallowed a frog?" he asked, using the humorous expression that made the question impersonal. He looked at the smoke-blackened rafters overhead.

Tseutsi laughed – a secretive chuckle. "Who is to say? Perhaps you should ask Sax ump ki if he has had any dealing with frogs."

The Kake began to cry and Tseutsi took him to his mother to be nursed.

From that time on, Cha it zit watched his mother. His status of approaching manhood forbade him close communication with her, but the more he observed her, the more he became convinced that she was going to bring a child into the family. He thought of how he would treat the baby. "If it is a boy, I will teach it. I'll help him learn. I'll play with him, too." The idea pleased him.

Then he resumed his work on the counter sticks. His one obsession was to become a great sla hal player. The game, where one gambled on the position of a plain bone, winning or losing valuable goods, held out the possibility of great wealth. He knew in his inner mind that he possessed a gambling spirit, *Tiolbax* by name, and that if it was properly treated, that spirit gift would help him win at sla hal. All he really needed, he believed, was a chance to show everyone how great a gambler he was.

So he whittled the long, slender sticks, which were used to keep count, against the day he could produce them and challenge guests to a game. His knife, of sharpened jade and dogfish sharkskin, served to shape and polish the cedar sticks. As he worked, he thought of the conversation he had had with his father a few days before.

"It is time to think of finding a wife for you," Sax ump ki had said. At first, the thought had shocked him. He had never thought of himself as a husband, much less a father. He glanced up from his work to watch his absent uncle's two wives and their children. He observed that the younger of the two women had finished nursing the Kake and was rubbing his naked round body with seal oil. She massaged his back and legs to make them strong and straight. His head, just recently released from the flattening board attached to his cradle board, was discolored and his eyes popped out a bit from the pressure.

"He isn't nice looking now," Cha it zit thought, "but his long, flat forehead will be attractive later." He felt his own aristocratic head. "It is better than being roundheaded like a low-class slave," he concluded.

He remembered the baby's cries when he was still confined to the cradle board, though, and wondered about the custom which no one questioned and which had come to the clan from distant, supernatural ancestors.

How would he feel when his *klootchman,* some day, did the same thing to his son. He choked the thought before it could seat itself in his mind. To question the traditional ways of the people was wrong. The great ones of the past were wise, half-mythical beings, and their inherited ways were sacred to the people. No, he would act as his clan always had.

Having controlled that errant thought, his mind returned to the thought of taking a wife. Then the face of the girl from Samish appeared as a vision. He savored her soft eyes, her gently rounded form, and a strange emotion engulfed him. Feelings swelled in his body which alarmed him, feelings foreign to him.

He dropped the stick he was carefully sanding and rushed to his sleeping place. There in the cold darkness behind the cattail mat that had been draped by his pallet, he shivered and felt the internal tumult subside.

"What is happening?" he wondered. Could the thought of the girl do that to him? He would think of her no more. Resolutely, he returned to the fire, and began to furiously polish the stick.

But he did think of her again. Whether he willed it or not, her face appeared, and with it the feeling, urging, even demanding, that he respond.

The next morning, Sax ump ki approached Cha it zit. "I have noticed that you are becoming a man," he began. "Feel your upper lip, there is a soft fur appearing on it. It is a sign of other things to come."

Cha it zit felt his lip and knew that it was true. "As you know, the young men pluck their face hair," his father continued.

The sight of Cha wentz painfully pulling his hair with a hinged clamshell had often impressed the boy.

"Someday, you will do this, too," Cha wentz had snarled at him through agony he dared not show.

"When you have aged, as I have," Sax ump ki said, "you will have earned the right to allow your fur to grow." He pulled at his long mustache, which drooped from both corners of his mouth. A sparse black growth covered his chin.

"Go to the beach and find an empty clamshell," Sax ump ki commanded. "Soon you will need it. Since you are becoming a man, it no longer is fit for you to share sleeping space with others of the family."

"But I have a mat to separate me," Cha it zit remonstrated.

"It is not enough. You will sleep in your own space beyond that now reserved for Cha wentz and his wife."

"If you wish," the boy answered unhappily. "I am being forced out," he said to himself.

He threw a bearskin robe around his body and ran out of the longhouse. The brightness dazzled and blinded him momentarily. Snow lay on the ground and a fiery sun burning in a blue sky tossed glittering rays with abandon to create bursts of light. It seemed that fires sparkled from everywhere. He ran in the snow, picked it up, ate it, tossed it, and when he was sure no one was looking, he rolled in it. His body became numb, but still he played.

At last he ran out onto the great tide flats. The salt water, frozen only where stream water had diluted it, lapped quietly at the icy shore. All was quiet, so very quiet.

Then he found a complete, open clamshell and raced back toward the longhouse.

He burst in. "You should see it," he called. "It is all white. The ground, the trees, it is white. Come and see."

"It is snow," Tseutsi said. "I have seen it before. It is better than the rain all the time. I will stay by the fire where it is warm."

Cha wentz ran out, tailed by his bride who followed him whenever he permitted it. Older children went too, running naked and barefoot in the snow.

Sax ump ki, Wanana, and his other wives, as well as his brother's wives, watched from the landing at the house front. Soon other children erupted from the village and joined the fun.

That night, Cha it zit took his sleeping mat, his bearskin robe, and his warm mountain goat wool blanket and moved to the other end of the longhouse. He hung his divider mat alongside that of Cha wentz, placed his few belongings beside him, and lay down.

It seemed strange, this new place. He was at the less prestigious end of the house, near the entrance and next to the area reserved for the slaves. He resented the move, but then, as he lay there, he began to rationalize.

"A man must earn his place. Now my place seems to be little better than that of slaves. Maybe that is all I am worth now, but I'll show them. I will be great. I'll win at gambling. Maybe I'll even become a shaman, as Tselique said I would. Someday I'll sleep on the platform at the head of the house. Yes," he promised himself, "I'll show them."

The time of mourning for Tselique had passed. Tseutsi's hair, now cut so that her ears showed, was beginning to grow longer. No mention was ever made of the old shaman's name or presence. Nothing of his remained in the house. It was as if he had never existed.

"He has gone to the world of the dead," Tseutsi had said, and that ended the matter.

But it was not over for Cha it zit. The legacy that Tselique had left him was always in his mind. The amulet he wore, a gift from Tselique, in a sack Latsi had made for him, was a constant reminder. When the others seemed to have forgotten, the boy remembered even more vividly the conversations he had had with Tselique.

The spirits, bringing their power, were returning to the village, as they did each winter. Strange things began to happen. Some of the people became ill and only recovered when their spirit songs came to be sung. Others became demented, struggling to handle forces rising within themselves. Fires were set in the village smokehouse and the people spent much time there, giving expression to the spirit power which they now possessed.

Cha it zit felt no great urge to visit the smokehouse. He heard the drums, and while they compelled others, they had no real meaning for him. His spirit power did not express itself. No song came to him that must be sung. He waited for the time to come that would have meaning – then his spirit would show: sla hal!

Cha wentz regarded his maturing brother with open disgust. Nightly he, along with other village warriors, and a number of headmen as well, disappeared into the smokehouse. All but a select group were barred from entering at that time.

Hideous yelps and frantic drumming vibrated from the huge building. Then there were times of absolute quiet.

"What is happening in there?" Cha it zit asked old Sta nek one evening when the noise was particularly exciting.

"Once I took part," he answered, his smoke-reddened eyes regarding the boy. He seemed lost in recollection. "We did some terrible, powerful things in those days. There was *squedelich* everywhere."

"Yes, but what?"

"If I were to tell you," Sta nek answered slowly, to let each word count, "they would kill you."

"Kill me?"

"No one outside of those who are initiated into their society must know what they do. I must never speak of it, and you must not know."

"Never ?"

"Do you wish to know badly enough to join them?"

"I don't know. How can I know if I don't know what they do?"

"That is the way it is."

The conversation was closed, Cha it zit knew, but his curiosity was pricked. How could he find out? Could he spy? Peek through the smoke hole? Maybe hide in the longhouse somewhere? The challenge intrigued him, and took his mind off the all-compelling subject of his changing body.

The next morning while the people were still drowsy in their blankets, Cha it zit rose and made his way to the smokehouse. He crept through its low entrance, expecting to find it empty. Instead, he came face to face with Cha wentz and a group of other warriors.

Cha wentz grabbed him viciously.

"What are you doing here, squirrel?" he asked.

"I'm just – "

"Speak. What are you doing here?"

"I wondered – "

"You wondered what we were doing. Squirrel, you do not belong here. There is nothing you should know. If ever you become a man, a real man like us, you may find out. Now get away from here."

Cha wentz pushed Cha it zit violently from the smokehouse.

The boy ran to the ledge above the water where he always found peace. It was cold, but there was a hint of warmth in the pale sun that showed through the sea fog.

Nothing seemed to be going the way it should. He had lost his place in the house near Sax ump ki. Now he was little better than the slaves. He was forcibly evicted from the smokehouse by his brother, and nothing was even happening there. All the family seemed to be involved in their own affairs. How he wished he could talk to Tselique. How he missed him.

Just then, the raven flew onto a nearby branch and cawed loudly. Cha it zit felt better.

"Oh," he said, "at least you are my friend."

The winter dances had taken a serious turn. The time for snatching new dancers had come. Those spirits without dancers to express them became violent in their need. Suddenly a dancer would grab one of the villagers and force him into the smokehouse. There he would undergo torment until his body submitted to the spirit and the song came bursting out.

Cha it zit knew all of that and he feared the dancers. As much as possible, after his encounter with Cha wentz at the smokehouse, he had quietly avoided any contact with him or the other warriors. But one evening Sta nek approached him as he was finishing his set of sla hal counters.

"You have not gone to the smokehouse," he said. "I hear that people say you are afraid to go. I do not believe that is so."

"I do not wish to be a dancer."

"When I saw you dance at your time of testing I knew you had a feeling for it. Come with me. We will show our faces tonight."

Cha it zit could not refuse, so he laid his work aside, picked up his blanket, and followed Sta nek.

When the pair entered, the dances were already in progress. Three new dancers, tightly wrapped in a cocoon of blankets, lay on the dirt floor near the roaring fire.

"It must be hot for them," Cha it zit worried.

Drummers sat together on mats at the head of the house. Singers, both men and women, stood behind them. Villagers sat on the platform that ranged around three sides of the building, leaving the entry open.

A lone dancer gyrated around the room. He seemed to be imitating an eagle, hopping and flapping his arms. At last he stumbled and fell from exhaustion. Before he had even hit the ground another leaped into the center and began his own dance. The drums changed cadence and the singers took up a new song. The dancer stopped by one of the blankets, shook it, blew into it, and stomped around it. A wail came from the captive inside of it. The blanket stirred, and a man, well into his prime, emerged.

Half crawling, he tried to stand erect. The dancer, grasping his arms, pulled him upright. Suddenly, there were others on the floor cajoling, pressing him to dance. An ecstasy seemed to come upon the man. He threw his head back and emitted a half scream, half wail. His legs jerked spasmodically and he danced. The others danced with him, imitating his movements, the drummers took up the beat, hitting their rawhide drums at the moment his feet hit the packed floor. He danced till he could dance no more and dropped, a quivering bundle of flesh and bones.

The dancers, now solicitous and careful, watched him till his movements subsided and he relaxed. Then they helped him to a mat where he rested. For him, it was over. He had become a dancer.

Cha it zit had watched the process of initiating a dancer before, but this time it became personal. He was reaching the time when it could happen to him. It was something he did not want. Being a dancer involved total commitment during the winter ceremonial season. Dancers lived apart during that time. They became dangerous to others, doing unspeakable things, even biting the flesh of people in their frenzy. One could not be a dancer and expect to spend time gambling.

It continued, other dancers felt the pull of their spirits and leaped to the floor to dance till they could move no more. The bundles on the floor remained still. No amount of coaxing could bring their spirits out of them. The dancers grew anxious. Those who watched knew of the dangers faced by the candidates if they couldn't respond. Strong men had been known to die or to go insane under such circumstances.

Now the dancers tried a new approach. In a group they danced around one of the prone, nearly suffocating initiates. They picked a bundle up and danced with him. Violent motions ripped the blanket open and a half-conscious man fell out. He crawled about on the floor, beating it with his fists. He drew himself up into a half crouch and flailed his arms wildly. Then he began to grunt. His grunt grew into a half chant, half song, and he danced. Again, the drums followed him. He seemed not to know where he was or what he was doing. The dancers kept close and guided him until he became aware. Then he fell, exhausted.

There was one left. No amount of cajoling or abuse could stir the person who lay there. At last, the dancers picked up the bundle, carried it to a cooler place, and appeared to forget it.

"What will happen to him?" Cha it zit asked.

"They will leave him there and try again later," Sta nek said. "He must dance to release his spirit. There is no other way."

"Will he die?"

"If the spirit doesn't come out, he will surely die," Sta nek said. "It is the way it is."

"Will no one give him water or food?"

"They will watch him. The dancers will stay with him, but he must suffer to make the spirit come."

The drums began again. Now anyone who felt the influence of power could dance. Far into the night the drums and singers would continue, as long as there was a dance needing to be expressed.

Children were sleeping at the back of the platform. Adults were tiring too.

"Let us go," Sta nek said. "My old bones are weary."

"Yes," Cha it zit said, gratefully. He had escaped the dancers for now, but he worried that his time would soon come.

The drums seemed to be in his ears. They throbbed through the night. Cha it zit slept fitfully. He felt he was suffocating, that he was the one in the blanket bundle.

When morning came he awoke feeling very tired. Wanana was already up, stirring her cooking fire. If Cha wentz had gone to bed at all, he would sleep half the day. He had been one of the dancers. Perhaps he will forget his girl-wife and stay at the smokehouse as the dancers usually did. That would please Cha it zit, for he had no wish to see his brother that day. It would take time for him to accept what he had witnessed the night before.

He aimlessly wandered out of the longhouse and let his feet lead him where they would. Morning sea fog lay along the water's edge, obliterating all but the shoreline. Overhead he heard the sound of geese calling as they flew.

"Spring is near," he realized. "Already our brothers are going toward the north. The spirits will go on their way to the farthest parts of the earth, on the air and under the sea." He was glad.

4

Sla Hal

The winter season was drawing to a close and there had been no opportunity for Cha it zit to play sla hal. His set of twenty counters and a pair of bones, plain male and decorated female, were finished and ready. He now spent time helping Sax ump ki make new fishing gear.

One day, while they were working together weaving twisted nettle stem twine into a fine net, Sax ump ki mentioned that the other village had agreed to the idea of building a fortification after consulting with their families.

"They asked me to lead the way," he informed Cha it zit. The boy noted that his father had, for the first time, confided in him. He felt that he had suddenly grown taller.

"They did well," he said.

"Now I must go to the village of Temxwigqsan and speak to the headmen there. If they agree to help, we will begin to work. I wish you to go with me."

Though Cha it zit's spirit leaped within him, he simply said, "Yes – yes."

"Good, we leave in the morning."

"Will Cha wentz go too?" the boy wanted to know.

"Of course; he is my first son, and as a warrior he will have much to say about our plans. A man's strength lies in his sons as he grows older, so a wise man includes them in his thinking and prepares them for the days to come."

Cha it zit concentrated on the knots he was tying in the tough nettle twine.

"Another thing," Sax ump ki continued, not seeming to notice his son's mental withdrawal, "there may be a sla hal game while we are there. Bring your sticks."

"Sla hal!" He forgot the respectful tone one must use when addressing an honored parent. "Sla Hal!" he repeated with joy.

The next morning he wrapped his sticks in a piece of old blanket, rolled his cattail sleeping mat, and took his bearskin robe. He asked Tseutsi for a few berry cakes and she handed him some from her store of food.

"It is getting low," she muttered.

Then he raced down to where Sax ump ki's large traveling canoe was tied high on the beach out of reach of the winter storms. Putting his shoulder to the stern, he tried to push it toward the water. It would not budge.

"The squirrel is trying to do a man's work," Cha wentz called, as he approached the beach. He was accompanied by two husky slaves who immediately bent their backs to the job.

The canoe slid reluctantly along the run that had been cleared of rocks and gravel and then slid easily into the water where it sat, gently rocking.

The slaves then lifted Cha wentz, waded into the icy water, and placed him in the canoe.

At that point, Sax ump ki arrived with his belongings and the slaves carried him to his place in the center of the canoe.

They turned toward Cha it zit, not knowing whether the boy had been awarded the status of his brother, uncertain of what to do. Cha it zit solved their dilemma by splashing out into the water and climbing into the bow of the canoe.

Cha wentz, with a strong stroke of his paddle, sent the canoe flying outward. Cha it zit used his own paddle to balance the powerful thrusts from the rear.

There was no sound except for the rush of water as the sleek dugout cut through the waves. The crescent bay with its beaches glistening golden in the morning sun was a rare sight at that time of the year. The trees growing above the beaches were bare of leaves, while at their backs, the evergreens and the alder stood sentinel above the village. Behind them, the delta of the small river lay barren and half submerged. The houses were fast disappearing as they rounded a projecting point, until at last they were lost to view.

The bay widened but they kept close to the shore. Ahead, cliffs loomed along the rocky beach topped by scrubby, windswept growth. They rounded a headland and were caught in a strong current. It seemed to pull the craft toward the shore. Cha it zit struggled to keep the nose pointed outward. He then felt the strain easing as Sax ump ki pulled with his paddle.

Land appeared ahead of them. "Shalahsheen," Sax ump ki called out. "At times one can run between it and the mainland. When the water runs out, it happens that way. Otherwise we will have to go around, or carry the canoe."

They approached the passage. Water was running through it.

"It is low," Sax ump ki stated, "but if we have the power to do it, we can make the other side before the channel dries."

He began a song, a power song to drive the canoe along. Cha it zit and Cha wentz joined him. The canoe felt it, and leaped across the water. By the time they had nosed into the deep blue water at the other end of the cut, only a trickle of water was entering the passage.

Now they were in the passage between the mainland, which had angled sharply to the right, and a large island known as *Samamao.* Ahead lay an inviting beach that curved to a rocky point. On they went, driven by a relentless Cha wentz. Cha it zit would have eased onto the beach and rested on its inviting sands, but his brother, his muscles honed and tense, would not stop.

They rounded the point to find another beach, this one even more inviting than the first. Above the sands were frames for drying salmon and for hanging sheltering mats on. A summer fish camp, Cha it zit surmised.

Yet another point and they saw the roofs of the village. Temxwigqsan was like no village that Cha it zit had ever seen. It was certainly nothing like his own. Swetquem consisted of five longhouses in a row along the water. This one had but two and they formed a right angle.

While he was marveling at the appearance of the village, several men ran out to meet them. Others were emerging from the houses. They carried spears.

Sax ump ki called out, "It is me, Sax ump ki from Swetquem, with my sons Cha wentz and Cha it zit. We have come to talk."

One of the men grabbed the bow as the canoe neared shore and others guided it to the beach. Then they beckoned and slaves came to carry them ashore.

Once inside the larger of the twin houses, the three were greeted by the headman. He was a big man – his belly hung in folds – but he was also well-muscled and he carried himself with authority.

Looking about, Cha it zit noticed that the shelves were lined with storage boxes, all full. "There is wealth here," he thought. He recalled Tseutsi's remark that morning and visualized the empty boxes stacked in their own longhouse.

"Welcome," the clan leader was saying. "When you have eaten and rested, we will listen to what you have come to say."

A woman spread mats for them to sit on near a brightly burning fire and then brought a wooden bucket of water and a ladle for them to drink from. She then busied herself about the hearth. She took a string of dried clams from a storage basket. After pulverizing them with a mortar and pestle, she tossed them into a wooden cooking box along with a handful of dried seaweed. Then, using wooden tongs, she picked up red-hot cooking rocks from beside the fire and dropped them into the box. Steam began to rise, bringing a wonderful aroma with it.

Cha it zit was suddenly hungry. His mouth watered as he waited for the chowder. When the woman brought the box to them, with more ladles, Cha it zit, Cha wentz, and Sax ump ki gratefully drank mouthfuls of the clam soup. They did not speak, as was mannerly, and kept their eyes on the food. When they were comfortably filled, they laid down their ladles and smacked their lips in appreciation.

Then the woman handed them shredded cedar bark to wipe their hands and faces, and removed the chowder.

"Now," said the host, "you shall tell me why you are here."

"Your hospitality is great. We are pleased," Sax ump ki began. "My two sons and I have come to ask your help in an undertaking that will benefit us all."

He outlined the plan he had devised for building a stockade of posts with a strong house inside for defense against the Yakultas and other fierce northern tribes. "All the Nuh Lummi people would be asked to contribute labor and warriors and all would benefit from the protection," he explained.

The headman squinted in thought. He considered. At last he spoke. "It sounds reasonable to me, but I must consult with others. You stay here for a few days. Then you will have your answer."

"It is as I had expected," Sax ump ki said. "We will stay, as you wish."

Each found his own activity. Cha wentz went to look over the land so he could be prepared to discuss the placing of the logs. Sax ump ki went with the headman to explain the plan to others. Cha it zit took out his gambling sticks, hoping that his host might suggest a game. He pretended to work on them, correcting a flaw here and there.

Children came to watch him work.

Then, several elders, men too old to do much work, approached him.

"I see you have made sla hal counters," one said, admiring the workmanship.

"I hope for a game," Cha it zit answered.

"Hey," another addressed him, "you are but a half-grown tadpole. Do you dare to bet against me!"

"I would be pleased to do so," Cha it zit replied.

"You are foolish, tadpole, to challenge a man that you know nothing about. And what have you to bet? How could you make it worthwhile for me, or anyone, to play against you?"

"I have my fur robe," the boy said. "It is a prime bearskin. It will keep its owner warm in any weather."

He held it up for the man to see. Wanana had cured it for him; she had rubbed it with oil till the hide was soft and she had kneaded it till it folded as a blanket. The black fur shone in the firelight.

"You shall have a match tonight," the old man cackled. "Tonight you shall shiver without your robe."

Satisfied, Cha it zit put his sticks away, folding them carefully into their blanket scrap. Then he went outside and hid them under a beach log. No one except himself should touch them until that night.

That accomplished, he wandered about the beach. The shape of the houses intrigued him. Seeing the old man he had spoken to earlier, engaged in conversation with several other stooped and ancient elders sitting on a log in the unseasonal sunshine, he approached them.

"So, has the tadpole changed his mind?" the older man greeted him with a toothless grin.

"No, but I wonder about this village. The double house is like nothing I have seen before."

"He can tell you about that," the old man pointed to a companion who was so withered and wrinkled that he looked like a skeleton with dried, brown seaweed draped over it.

The emaciated one jerked his head and looked at Cha it zit with milky eyes. He pointed a long claw toward the houses and spoke in a childlike voice.

"Yes, I know about the house. It is called Lulumis. All the people call themselves Nu lummis now. They are named for another house, built like this one a long time ago."

Cha it zit listened with respect, but he doubted that the ancient man was telling him the truth. One did not insult elders, especially one that old. "He has the sickness of age," the boy decided. "I will humor him in his fantasies."

"I do not understand how that could be, grandfather," he replied.

"You do not know our history. Young people today do not care about our ancestors and the great things they did. Since you asked, I will tell you."

Cha it zit slid down onto the beach sand, his back resting against a log comfortably, expecting a long recitation.

"Our people were the first people," he said. "The first man dropped from the sky at the northeast end of Xulelq Island. He was the ancestor of the Klalakamish people, as they called themselves then. Then a terrible plague came upon the people and almost all of them died – all but the last man. He gave his house to a man who already had one at a place called Flat Point on Sweluc Island. The space was small, so he couldn't put them in a line to join them together, so he set them like this." The old man bent his elbow at an angle to show how they looked. "They called the house Lulumis which, as you know, means facing each other. Then, when the people left the islands to come here, they moved the houses here."

"That house is very old," Cha it zit observed.

"It is old. It was old when I was born," the old man cackled. "That makes it very old, for I am the oldest of the elders."

"Was there really a sickness that killed all the people?"

"A terrible sickness. No one knew why it came. We had offended a spirit, we knew. Our shamen worked to find the cure, but in the end they died, too. It was told to me by my grandfather when I was a boy. His face was full of holes. He said the sickness left him like that."

"What else did he tell you about the angry spirit?" Cha it zit was intrigued by the story.

"Well," the ancient scratched his thin, white hair, searching for lost memories. "I seem to recall that sores broke out all over the people's bodies. They became hot as fire, and yet they shivered with cold. They spoke words we could not understand. If they lived, as my grandfather did, then their faces were full of holes."

"What did the shamen do?"

"They tried all their powers. Nothing helped."

Cha it zit thought much about this. Suppose the sickness came back. It only went away after most of the people died. Now that there were more people, could it return? With Tselique gone, who could help them? He had been the greatest of the shamen.

"Were there a lot of people then?" he asked.

"More than grains of sand on the beaches, I was told. Once we were many. Now we are few. Once one village held more people than all of ours put together now."

"How terrible it must have been," Cha it zit said sadly.

"It is growing late," the toothless one interrupted. "My old bones feel the chill."

It was truc; the sun was nearing the top of Samamao Island across the channel. Soon darkness would come.

"Warm your sla hal bones, tadpole," the old man chuckled, "and bring the bearskin. I will sleep warmly tonight."

Cha it zit watched the old man rise laboriously and shuffle toward the low entrance of the double houses. Then he dug his bundle of sla hal equipment from beneath the beach log and walked slowly across the beach away from the village.

As he walked, he extracted the playing bones and held them, one in each fist. He concentrated on feeling the bones, the female with its decoration of the incised lines and circles, and the plain male. They fit neatly in his palm and felt smooth as river-polished stone.

"Tonight you must know the power of the Tiolbax spirit," he told them. "That power was granted to me at the time of my spirit quest. Now it will come to you through the same spirit."

A wondrous thought came to Cha it zit as he walked, not really seeing where his feet were taking him. He realized that he was beside the channel he had come through on his spirit quest. Beyond it, so close that he could see each tree that fringed the shore, was the island of Samamao. He recalled that the channel was the place where he had received his vision. He remembered how it had happened. His canoe had grounded on the submerged roof boards of a longhouse and he had heard a voice from the sea saying, "I am Tiolbax. It is through your house and my help that you will gain wealth and prestige."

"Could it be that this is the place where my own house will stand? Here, near the village of Temxwigqsan?" Cha it zit wondered.

The evening was upon him and cold was settling over the land. The sla hal bones grew warm in his hand, they seemed to radiate heat. "The bones have a life of their own," he thought, "and they are warming me with their vitality."

Satisfied that his spirit power was with him and that he need not fear to play sla hal with more experienced players, who, no doubt, had power of their own, he turned toward the longhouses.

It was quite dark inside, but the flickering fires cast out ripples of light, enough to see that the elders were preparing for the promised sla hal game. First they placed two planks a man's length apart. Then cattail mats were laid between them. Along the plank, at intervals the distance of a man's arm, they laid stout beating sticks. That done, they went to their storage areas to gather up items to wager on their skill as players.

Cha it zit placed his counting sticks between the planks, and then placed his bearskin robe on the mat in front of them. He kept his bones in one hand lest someone touch them and interfere with their power. That done, he sat down behind his sticks and waited.

He had not seen Sax ump ki or Cha wentz all day. Suddenly he felt lonely and very vulnerable. He wished that they would appear. He took some comfort from the warmly reassuring bones in his hands.

The old man arrived and placed his belongings on the ground before his place, opposite Cha it zit. Then the others came and, arranging their possessions before their places on either side of the ancient one, they too, sat.

Others arrived and ranged themselves down the board. Cha it zit sat alone. He became alarmed. He hadn't thought about the fact that he needed a team to support him. Then there was movement behind him, and he saw with gratitude that Cha wentz and Sax ump ki had joined him. There still weren't enough players, but three was better than one. Then, several old women, finding no places on the opposing side, decided to chance playing with the untried boy. A few others, those who were not acceptable to the experienced team Cha it zit faced, took places alongside the women and the game began.

The old man placed his bet, a newly made wooden storage box of fine workmanship, in front of himself and directly across from where Cha it zit sat. The boy nodded his approval. Then the beaters took up their sticks and began to hit the boards with a firm, steady beat. The players began to sing the sla hal song, the thin, nasal soprano voices of the old women dominated the throaty bass of the men. The steady beating of the sticks and the magic of the repetitious singing stirred Cha it zit. It was as if the song had entered his body. He swayed to its tempo.

As a guest, he had first chance to guess for the position of his opponent's unmarked male bone. The elder, after showing his well-worn pair of male and female bones, hid them in his fists under his blanket. He made a show, moving his hands to the beat of the tapping sticks, back and forth beneath the blanket, and then bringing his closed fists, fingers down, into view. He swung his arms rapidly to the left and to the right, always keeping time with the pounding of the sticks.

Cha it zit, following an impulse, guessed for the plain bone to be in his opponent's right hand. As was customary, he slapped his breast with his left hand to show he had made a decision, and pointed to the old man's right hand with the forefinger of his own right hand. All of his movements he matched with the steady beat of the sticks and the chant of the singers.

The guess was correct and the old man tossed his own playing bones to Cha it zit, who then took one tally stick and stuck it in the ground in front of himself. His co-players sang out with enthusiasm over the first win. Cha it zit then gave his opponent's bones to Sax ump ki, including him in the play. Now father and son hid their bones as the old man had done, and the game continued, growing in complexity and excitement. Occasionally the old man won his gambling bones back from Sax ump ki, along with a tally stick, but never could he guess correctly for Cha it zit's plain bone, and the boy's guesses seemed to always be lucky.

The people watching began to murmur among themselves over the extraordinary success of the young upstart. "He has control over the bones," someone close behind him observed.

"Truly, he has a strong spirit helper," another replied.

The hours passed. Some observers, encouraged by Cha it zit's success as the number of tally sticks standing in front of him grew larger, placed small bets of their own on the mats which separated the players. Then, finally, all the tally sticks were lined up in front of Cha it zit. He had won the first game.

Then he took the box that the old man had bet and placed it on top of his robe. Since no one on the old man's team had moved to match the smaller bets, they were left in place awaiting the next game.

Next, the old man showed a blanket, somewhat worn, but a worthy bet. Others of his team, expressing loyalty to their leader, added to the bet until it equalled Cha it zit's offering of both the box and the robe.

The counting sticks were again placed in the center, and again the sla hal song and the beating of the sticks, and again Cha it zit won the goods. The playing continued until the pile of goods in front of Cha it zit threatened to hide him.

"It is enough," Sax ump ki said gruffly to his son. "We came as friends. The temper of our host is beginning to show."

"One more game," pleaded the excited boy.

At that point, the headman, who had been watching, rose and tapped the elder on his shoulder.

"Grandfather," he said, "let me take your place. Let me show this tadpole, as you called him, how a great sla hal player wins."

The old man, stripped of his few possessions, had no choice but to give his place to another.

Cha it zit watched the huge man ease himself to the floor. Then he saw him toss a stick of wood onto the mat. "This is my canoe," he said. "It is equal in value to the goods you have won. It is new and it is a fine craft. We will play this one game. All you possess for my canoe."

"It is as you wish," Cha it zit replied.

"Have a care," Sax ump ki whispered. "If you win we may make an enemy."

But Cha it zit gave him no heed. The game possessed him. The headman was good, the boy observed. He handled the bones with a sure, deft touch. Every sense the boy possessed, he now brought into the play. He watched the eyes, the mouth twitches, he even felt, in his mind, the tightening of his opponent's muscles.

On and on the game went. The tally sticks moved back and forth. Sometimes the host and his team was ahead, and then Cha it zit and his players held the most sticks. At last the boy began to win. Mentally, he called on his spirit power and he felt it respond. Finally the game was over.

The opposing team departed, disgruntled. Cha it zit distributed part of his winnings to the beaters and part to his father and Cha wentz. A large pile remained. He motioned to those who had bet on his skill to take a share from the winnings.

He spotted the old blanket. "The ancient one will be cold tonight," he thought. He handed it to one of the old women who had joined his team. "Here, grandmother," he said, "take it to the old man. Tell him he needs it more than I do." She hobbled off to do the errand.

Cha it zit gathered up his winnings. He placed as many of the small articles – fish hooks, line, tools, and a carved pipe – in the box. The rest he placed in a bundle which he stored in a far corner of the longhouse.

The host approached him. "My canoe is on the beach, the first you will come to," he said as he turned to go. Then, suddenly, he stopped and took Cha it zit roughly by the shoulder.

"Tell me how you did it," he demanded.

"I have the power," Cha it zit answered simply. "I received it on my spirit quest."

"They called you tadpole," the host observed. "I think they were wrong, they should call you salamander."

Cha it zit thought before replying. The man was paying him a compliment, he knew. A salamander has special powers. His answer must be wise, but how could he speak to this man who had just lost a prized canoe? He decided to change the subject.

"Your village is very pleasant," he said, "but it must also be a dangerous place to live. Do the northern people bother you here?"

At the mention, the big man tensed and his eyes seemed to shoot out fire. "They come in the night and steal our women and children. They come in the day and kill our people. We fight them, but they are too powerful for us."

"Then you will help us, as my father asks?"

"Yes, we will help," he said. "Salamander, I would rather be at the side of one like you than against you."

"Then your canoe may yet be of service to you," Cha it zit answered.

The night was almost over. Already a glow in the east announced the dawning of a new day. Suddenly Cha it zit was very tired. He wrapped himself in his bearskin robe and slept. No one disturbed him, for instead of enmity, his exploit had brought him respect. When at last he awoke, he hurried to the beach to find his canoe. It was all the headman had said it would be. As he was examining it the man, himself, appeared.

"See how sleek it is," he said, caressing it lovingly with his huge, fat hand. "It will carry you safely and swiftly to wherever you want to go."

"Yes, I can see that it is a great canoe," Cha it zit said.

"You will be a great man yourself, if you continue to play sla hal like you did last night."

"I will," Cha it zit assured him. "It is my gift to win."

"Ah, yes. And you will have many wives, eh?"

At this, Cha it zit dug his toe into the sand, watching as he stirred the soft, fine powder.

"I embarrass you. You are young, and no doubt have not known a woman. But," he chuckled, "remember, when you think about it, that I have a daughter who will soon be of a marriageable age. It would be a good thing to join our villages in a liaison such as one between you and her would be."

"My father has spoken to me of marriage. He has said that he will chose who she will be. You must speak to him."

"You are right," the headman agreed quickly, "I spoke out of turn." Then he playfully poked Cha it zit in the stomach and said slyly, "Just the same, think about it. You have some influence with your father."

Later on, when Cha it zit had the opportunity, he mentioned, in a casual way, the conversation to his father.

Sax ump ki was angry, "He is a conniving old fool," he said. "But we have to treat him with care, the success of the stockade depends on his cooperation."

"I know," Cha it zit said, "that is what worries me."

Cha wentz was satisfied with the location for the fort. The people of the village had agreed to provide the land and manpower. Now all that remained was to enlist the aid of other Lummi villages and to build it.

"It is time for us to go," Sax ump ki announced to the people of Temxwigqsan. "Your hospitality has been great. Soon I will return with others to begin the to work on the fort."

Cha it zit gathered his new-won wealth and carried it to the great canoe. While he was arranging the goods inside so as to balance the weight and protect it from sea spray, the ancient sla hal player arrived.

"I was wrong," he began, "you are no half-grown frog. Others have called you a salamander, one who possesses squedelich, magic powers. I think that is a better name for you."

"You could not know, old one," Cha it zit answered kindly. "I did not like to take your possessions, you have little time left to enjoy them, but the power I had would not be denied."

"I will remember the gift of the blanket, and I will remember you. When we meet again we will speak of things you wish to know."

Cha it zit wondered what the old man meant, but there was no time to ask. It had been decided that the new canoe would be towed behind Sax ump ki's traveling canoe. The water was calm, so there would be no difficulty, and, with the three paddling, they should make reasonably good progress.

The drag of the heavily loaded canoe slowed the craft paddled by Cha it zit, Sax ump ki, and Cha wentz. By the time the sun was at its highest, their seasoned arms ached from the strain. True to their tradition as warriors, Cha wentz and Sax ump ki tried to over-ride the pain and their cramping muscles. Cha it zit, his body not yet at full maturity, called upon all the resources that he had obtained during his spirit quest training.

"I will not submit to the ache," he told himself. He would have beached at one of the coves they passed if he had been alone, but he could not suggest it to his father and brother. He imagined their reply: "What a puppy you are."

At last the sight of their village greeted his eyes. Already the people had seen them and had gathered at the water's edge. Latsi was among them, in the forefront as usual, Cha it zit noted with a feeling of pleasure. He missed their former camaraderie, their shared confidences, their serious talks. In his new stature of almost-man, such easy relationships with a girl or woman, even his sister, while unacceptable, was still second nature. There was so much he would tell her if he could; his success at sla hal, about the strange house called "Lu lumis," the offer of his daughter by the headman.

While the two male slaves leaped into the water to grab the prow of Sax ump ki's canoe and pull it to the shore, the villagers ran out to Cha it zit's new canoe and untied it from the tow rope. Cha it zit heard the people saying:

"Where did it come from?"

"What a fine canoe."

"Feel the sides, how true they are."

"No one suspects it is mine," he laughed to himself. "I will let my father tell them about it. That way I can watch their faces."

After the slaves had carried Sax ump ki and Cha wentz ashore, they turned to the new canoe, to pull it high on the beach and unload it. When the contents had been placed in Cha it zit's part of the longhouse, the people repeated their questions.

"Later," Sax ump ki called out over his shoulder as he hurried to report to the other village leaders about his talks at Temxwigqsan.

It wasn't until that evening, when all the villagers had gathered together in the house of Sax ump ki, that they were told the details of the plan to build a stockade and of Cha it zit's exploits.

"He has great potential," they agreed. "Someday he will be wealthy and give many potlatches."

"Yes," he concurred.

5

Stranger

Cha it zit sought Sta nek. Ever since his return from Temxwigqsan he had thought about those people, particularly the headman who had offered his daughter.

Spring was soon to come; there was a pleasant softness in the air. Sta nek was sitting on the beach sand, leaning against a log, a knife and partly worked slab of cedar lying at his side. The old man's eyes were half closed, a look of contentment on his face. The boy was reluctant to disturb him, so he sat on the log and waited.

"Umpff," the old man snorted and roused himself. "You?" he questioned, his mind befuddled for the moment.

"I would talk to you, grandfather."

"Yes—yes, what would you say?"

"Do you know the house called Lu lumis in the village of Temxwigqsan?"

"Of course," he answered indignantly. "Why wouldn't I know a place so close by. One of my wives, long departed, came from there."

"Then you know the headman? How would you describe him? Is he an honorable man?"

"He is a clever man. He is tricky, but he means no harm."

"Have you seen his daughter?"

"Oh – ho! Now I begin to know what you are talking about. It is his daughter that interests you," he cackled.

Cha it zit was embarrassed. He would have turned his back on Sta nek, but we wanted to find out what he could about the prospect of such a marriage as the headman had proposed. So he sat and waited until the old man stopped laughing.

"If you want to know what happened long ago, I can tell you, but about his daughter I do not know."

Cha it zit's interest was aroused. "What happened long ago?" he asked.

"I am thinking of our ancestors, those of all the Nuh lummi villages, that came from the islands where we now go to fish and dig clams and camas. But they came from different villages. I will try to remember what has been told to me." He pondered, then continued, "Some of the people came from the island we call Skwamana. They were the Klalakamish people. They now live at the village of Temxwigqsan which you are interested in. Others came from the island Swelax and they were called Swelax people. Our own village came from the part of that island where the sun sets. We called ourselves Allulung people. You see, we are now all called Nuh Lummis. We have different ancestors, so a marriage between us is acceptable.

"Now," he continued, "I am reminded of a story that was told me when I was a boy. We were not the first people to live in our present villages. There were others here before us. They called themselves the Skalakhan people. Their nets and weirs were in the same places we put ours now in the river we call Lummi."

Sta nek stopped to scratch his head and poke at a small, shell crab with a stick. "I'm not sure how it happened, but a young man of the Swelax people got the spirit power of a warrior. I seem to remember that he dove in a lake on the island near his home for the power. He was able with the help of others to defeat the Skalakhans and they agreed to move up the river to save their lives. That was long ago, but some of those people still live there. We do not see them, for they still resent losing their tidelands and fishing places."

"I don't understand." Cha it zit looked puzzled. "We do not own the land as one owns a dog or canoe, or even a house. How can people lose their land?"

"No, we don't possess the land," Sta nek explained. "We only possess the right to use what the land and the sea give to us. Our family has the right to fish at a certain place, or to harvest ferns or camas in a certain spot. All the animals and birds and people have a right to use what is here. But no one owns the mountains or the ocean or the rivers or the lakes. They are gifts for every creature to use."

"Grandfather," Cha it zit said after having considered the words Sta nek had spoken, "you have told me of our people, now I am wondering, do you think there are others besides us?"

"What do you mean? Why do you ask?"

"I have been told of strange people with white skin who came on a spirit canoe to this place long ago. Are there such beings?"

"I, too, have heard of them. Perhaps. Perhaps. I do not know. There are many spirits, some good, some bad. Perhaps the beings were spirits."

"What have you heard?"

"Not long before you were born," he began, "some of our people went to the place we came from, Allulung, to gather clams and fish. It was early in the spring about this time. The winter had been harsh and they needed food. A canoe, larger than any of the largest war canoes of the

Yakultas, with huge white wings, appeared before them. Then, strange creatures with white skin, wearing an abundance of clothing fashioned in a most peculiar manner, got into big fat canoes. They moved them in a most unusual way. They seemed to be pulling them with long sticks."

"They sound like the same ones I was told about," Cha it zit observed, careful not to mention that it was Tselique who had narrated the story.

"The men were gone, trolling for salmon, so the women were at the campsite alone. They were cutting nice fat halibut for drying when the strange beings approached them. They spoke in an unusual language, but the women realized they wanted food. They were frightened and gave the spirits, or whatever they were, the halibut. Then the beings handed the women some strange beads. I have not seen them, but I heard that the beads were like the sky, blue, and you could see light through them."

"I wonder what they were?"

"We may never know."

"Has anyone else seen the beings?"

"Another time, not far from here," Sta nek said. "When you went to Tẹmxwigqsan you went past Shalahsheen, the place where the water runs through the land, and then dries up when the tide goes out."

"Yes, I remember."

"Long ago, even before I was born, according to the story, some men beings came in a canoe such as I described and landed near Shalahsheen. They had a strange box which they carried ashore. It was of wood, but it was not square; it was round like a basket. They made a great noise, so that the people could hear them. When the warriors came, the beings pulled a plug out of the box and water came out. They collected the water in a sort of ladle and drank some. The more they drank, the stranger they became. They sang strange songs and danced most unusual dances. Then they passed the ladle to the warriors. The warriors, too, were affected in a most unusual way. Soon they were all laughing and acting as if they were good friends, dancing and singing."

Cha it zit snickered at the mental picture his mind was conjuring up. "After that the men brought out some pelts of beaver and sea otter. They indicated that they would trade more of the contents of the box for furs."

"Did they?"

"The warriors were not hunters and they had only a few poor skins, so the men took their box and went away."

"How strange," Cha it zit murmured.

Just then Cha it zit noticed several canoes rounding the point. Yakultas?

"Look," he pointed.

"My old eyes deceive me," Sta nek said.

"Three canoes coming this way." The boy rose to call the alarm. Several villagers who had been repairing a canoe on the beach ran to warn of impending danger.

The canoes grew until their occupants were clearly visible.

"That is What la cum!" someone shouted.

"He has brought others with him. Call Sax ump ki. Tell him his brother has returned," another added his voice to the general uproar.

By the time the canoes had reached the beach, a group had assembled. Sax ump ki stood on high ground. His brother's two wives and their children stood some distance apart.

Sta nek had risen, painfully, and was hobbling toward the canoe run. He was obviously anxious to greet his younger son and to see who was accompanying him.

Soon the slaves had pulled the craft close to the beach and were assisting the party to land. What la cum was carried first. Then a woman, who rode ashore on the back of a husky slave. Others followed, several young men and a girl, just beginning to show the roundness of womanhood.

"Welcome back, brother," Sax ump ki called.

"Who are you bringing to us?" Sta nek queried, a trifle fretfully.

The canoes still sat low in the water, the sign of a heavy load nestled inside. What la cum's waiting wives and children kept a respectful distance and made no sounds.

"In good time – in good time," What la cum said, smiling so broadly that his face seemed to split from side to side.

The group made its way up the beach, Sax ump ki in the lead. They entered the longhouse and then Sax ump ki halted them.

"Brother," he said, "you have brought guests, and as head of this house, I welcome them. Our women will prepare food. Meantime, they may rest, for you must be tired. Now, brother, I would speak to you."

The two walked to the head of the longhouse where Sax ump ki had his quarters. They sat on the platform and were soon deep in conversation.

Cha it zit had no wish to spy but he was curious about their discussion. He was about to inch as close as he could and still be respectful when Tseutsi called to him.

"We need water from the creek."

Obediently, he picked up two wooden box buckets and carried them out of the house. There by the clear, bubbling stream, he met Latsi, who was already filling several water buckets.

"Who do you suppose they are?" he asked.

"Don't you know?" Latsi challenged him, an impish grin on her face.

"I know only that it is our uncle. That he has been gone for a long time, and that he returns with a woman, a girl, and some warriors," he answered defensively.

"Oh, Oogli," she answered, using his old pet name, "don't let's quarrel. I was only teasing you. I see you so seldom – for a minute it was like the old days – before you became almost a man."

"I'm sorry, Latsi. Living in a man's world makes me feel – well, different. But who are they?"

"The woman must be a new wife. I heard his other wives talking. They wanted him to find another woman to help with the work." Then she giggled, "I think he liked the idea too."

"What about the others?"

"Silly. Don't you know anything?"

"Latsi – I have had many things to think about."

"Well, the girl is probably her slave. She is a high-class woman, you can tell. She would never go without someone to help her. You know, do her hair, help her make new clothes and baskets."

"And the warriors?"

"Her brothers, probably. They will stay here for a while to see that she is cared for. If anyone misuses her they will have to account to them."

"I see," Cha it zit answered. But he didn't really understand it all. When Cha wentz took his wife there was a lot of protocol to observe and discussion over her dowry and the privileges she would bring to Cha wentz. His uncle had simply gone to find a bride. Why? What was the difference? Not wanting to show his ignorance any more than he already had, the boy filled his buckets in silence. Latsi had already left, so he walked slowly back to the village alone.

When Cha it zit had set the water buckets beside Tseutsi's fire, he slid into the shadows of the sleeping platforms. There, from his own area, now piled high with his newly won possessions, he was able to observe the newcomers. The woman, he could tell, as Latsi had said, was of a high caste. She bore herself with pride, and her head had been shaped in the elongated form as only the nobility could do. She was not young, but neither was What la cum. "Maybe she has been married before," he thought. "That could be why What la cum got her so easily. Or maybe she cannot bear children. Such a wife is not worth much."

Then he studied the young men. "If they are her brothers, they are much younger than she is. Or are they her sons? No," he decided, "the sons would stay with their father's people."

The slave girl was making herself useful. She was tending the several toddlers of What la cum's other wives, keeping them away from the cooking fires, cajoling them into a child's game of "find the yarn ball." Her head was round, the shape of a slave's or low-class person's head. It appeared crude and ugly to Cha it zit, although he had to admit that she had nice features. Her hair was healthy and shone in the light of the fire. Her cheekbones were high. Her eyebrows were thick and formed raven's wings about her dark and somewhat slanting eyes. He noticed how strong her body was. "It's from hard work," he reasoned, "but she isn't too muscular, just strong. She must be about my age, not as young as Latsi."

He wondered where they came from. There was something very different about them, but Cha it zit couldn't think of what it was.

Tseutsi looked around, scanning the people in the house. More had gathered to see What la cum and to get a look at his guests, if that's what they were.

"She is looking for me," Cha it zit said to himself. "I know that look. She needs something." So he left his place and went to help her. In doing so, he passed close to the slave girl, who was tagging after the little children.

She looked at him for a moment and then dropped her eyes demurely. There was something about that glance that stirred the boy. He felt an empathy for the girl. "I wonder what it is like to be a slave?" It was a strange thought, one that had never occurred to him before. Slaves were a part of life. They just were, if you were wealthy enough to afford to buy them. If you weren't, you went out and captured one if you could. They were possessions. Possessions didn't have feelings. But, still, he wondered.

At the evening meal, the new woman served What la cum. He sat apart from the rest, eating with no sign of emotion, pleasure, or displeasure. When it was over, she cleaned up the utensils and debris, then sat beside him on his mat.

That was when Sax ump ki mounted his platform and called the family, including relatives from the other houses, to listen to his words.

"Tonight we welcome What la cum's new wife. She comes to us from the S'klallum people who live far to the south of us along the beaches of the great strait that leads to the father of waters, the endless sea.

Sax ump ki gestured and What la cum led the newcomer into the circle of firelight. "I am Xedewa," she said in words with a foreign accent. "My husband's people will be my people."

"You have heard her words," What la cum said in an almost pleading tone. "Accept her as one of us. Her life has been difficult. Let us help her to feel that she is home."

Tseutsi pulled herself up from her sitting mat and hobbled toward Xedewa. She put her withered hand on the woman's arm and smiled a toothless grin. "You shall be as my daughter, the same as the other wives of my sons."

"Thank you, mother." Xedewa was visibly touched. "I am only a small wife, I bring only a small gift, and there are others of seniority; but I will do your bidding. I will work hard."

"My new wife is of noble lineage," What la cum spoke on her behalf. "She has had great grief, and life has been hard for her. If her bride gifts are small, it is not because she is of a low family. We will speak of it no more."

"My goods are in the canoe," the woman said. "I ask my nephews, who accompanied me, to bring them so that you may see what I add to your household."

The two young men left quickly to do her bidding. Others came forward to greet the new wife. Wanana took Xedewa's hands and laid them to her cheeks. Cha it zit nodded his head in recognition, as did Cha wentz. Then old Sta nek spoke to her.

"My daughter, you and those who came with you fill my heart. Our house was empty of all but a few after the great sickness. Then the raids of the Yakultas took more. Now I see it filling again. It will once more be a great house. May you bring us many babies to give us joy."

Xedewa smiled at his words. Her nephews carried in bound boxes and bundles and laid them at her feet. At a gesture from her, they began to untie the twine and unroll the mat wrappings. She then displayed carefully made baskets of a design foreign to the people. They murmured over them.

"We call them *twana baskets,*" she explained. "If you wish, I'll show you how they are made and what the designs mean."

The young men held up several white blankets with dark brown and gold designs woven into them. There were bentwood boxes, trays, and ladles – all of a high quality of workmanship.

"These are the gifts of my people," she said simply.

"We accept," Sax ump ki said. "At a later time we will visit your people and give them gifts in return."

"Now, it is customary for a husband to enjoy the company of a new wife. There is space in the house for you. Place your mats for privacy. There you will remain."

Xedewa turned to go, What la cum following.

Sax ump ki stayed in his place and the people sat quietly, feeling more was to follow. Then Sax ump ki beckoned to Cha wentz, who joined him. "We face difficult times," he said. "Our food supplies are low. Spring shows her face and then disappears in rain and sea fog. The time for food gathering is not yet, but we must begin to hunt and fish soon, or we will be hungry."

The people murmured.

"Another problem arises," Sax ump ki continued. "The Yakultas haven't come back yet, but they will, and they will bring more warriors even than before. It is necessary for us to begin to build the stronghold at Temxwigqsan. Others will join us, but even so, it will take the strong young men from our village to lead the effort."

Tseutsi grunted. "Food. We must have food. That comes first."

Sax ump ki stared at her. "We must have both food and protection. Dead people don't need food, mother."

"My son," she pointed a finger at him, "I have sat and listened to war talk, and then I have dug clams till my back ached to feed the warriors. I have not spoken. Now I do. We have babies, the children of your wives and your brother's wives. We must feed them, too. I say food first, then the fort, if you must build it."

Sax ump ki's face reddened. The veins stood out on his neck. No one had dared oppose him so openly before. "Woman," he shouted, "I am not a fool. I say we will do both. Now listen to my plan."

Tseutsi's eyes hardened and her mouth tightened, but she held her peace.

"We will divide our labors," Sax ump ki continued. "Cha wentz, as a warrior and my oldest son, will take the responsibility of the building. He will select the site, which in fact, has already been accomplished, lay out the plans, and supervise the workers. I am asking that our guests from the S'klallum people, nephews of my brother's new wife, remain to help. I will spend as much time as I can on the work, too."

Tseutsi snorted and was heard to mumble, "Men!"

"My brother and Cha it zit will fish for us. They will net and troll in our usual places. Those women who are able will visit the camas beds and clam beds and lay in a store. Tseutsi and Sta nek will remain here, as usual, to care for the babies. Wanana is with child. To go or stay will be her decision. Latsi is becoming a woman – her time of confinement may be soon. That worries me, because she will be needed to work."

"I will go," Latsi said.

"So be it," answered Sax ump ki, "but we are risking much – the contamination could ruin our food supply."

"Then I will hide," she answered.

"I go, too," Wanana said.

"There is much to do," Sax ump ki concluded. "Tomorrow the building begins."

6

Spring Comes

In spite of Sax ump ki's firm resolve, the work on the fortification did not begin the next day.

Cha it zit was awakened before dawn by a raucous noise. It filled the quiet air with an insistent, wordless chatter.

"Ducks," he shouted. "The ducks are returning."

Instantly, the household was alert. Cha wentz and Sax ump ki grabbed the large net they had just finished repairing. Cha it zit found the long poles where they had been laid alongside the house. What la cum emerged from behind Xedewa's newly hung mats looking sleepy and befuddled. Even old Sta nek rustled about looking for the duck decoys.

By the time the first light of the spring sun had penetrated the darkness, the family was on a slope of grassy meadow above the beach. The men tied the net to the two long poles, and after setting the butts in special holes, they raised it into the air. The women brought beach rocks to pile around the poles to hold them securely.

Then the decoys, carved cedar ducks, were floated in the water below, secure on tethers of nettle twine.

"We will feast tonight," they rejoiced. After a winter of eating dried salmon, dried fish eggs, dried berry cakes, and dried clams, fresh barbecued duck was appealing.

They watched as the first ducks spotted the decoys and dropped toward the water to join them, only to be trapped in the net. As fast as they would snare themselves, the men clubbed the ducks and removed them from the net.

Already the women were digging a fire pit and gathering beach wood. Using fine cedar twigs as tinder and coals brought from a smoldering cooking fire, they coaxed a flame in the pit. Bit by bit it spread until there was a trench of brightly burning flames. These they fed with wood of the alder tree. When the wood had been reduced to glowing coals, the fire would be ready. Meantime, the pile of slain ducks was growing.

The sun was high in the sky, riding between puffs of fleecy clouds, when the ducks were placed on spits to cook. There was the pungent smell of burning feathers at first, and then the delicious aroma of cooking flesh. The people's mouths watered as they waited.

"Cha wentz," Sax ump ki called, "come." He was sitting on a log below and some distance from the net where the beach sand was still smooth and wet from the retreating tide.

After Cha wentz joined him he called to Cha it zit. "You might as well know this, too." Glad to be included, the boy took his place beside them.

Sax ump ki drew in the sand with his forefinger. "This is the village Temxwigqsan." He drew a square. "The stockade should enclose the village." He drew a rectangle around the village. "Here in the center, raise a huge pole so that a pitch-wood torch can be attached to the top. That way the village can have light in case of an attack."

"It will be done," Cha wentz assured his father.

"Now," Sax ump ki continued, "I have given much thought to the barrier. Here is what I have decided. First, cut cedar logs. Many tall, stout logs will be needed. Each log should then be split in half lengthwise. After that, each log should be sharpened at one end into a wedge shape. Lay the logs evenly in rows about double the span of a man's arms. That done, cut a groove in another log in a wedge shape and fit it over the shaped points so they are held tightly together. Such sections, implanted side by side, braced from the inside by other logs, will form a protective wall that can be raised, lowered, or even moved, if necessary."

"It will be strong, too," Cha it zit interjected. Cha wentz stared at him, showing resentment at his interference. Undaunted, Cha it zit continued. "I am wondering how big the barricade will be."

Sax ump ki smiled. Two eager sons were a blessing to a man, and his twinkling eyes betrayed his inner delight.

"You are thinking that the village will grow."

"Yes," Cha it zit replied, sober and noncommittal lest his own reasons for asking became known.

"We will make it big enough, but think of the type of construction I suggest."

"I see." Relief in his voice. "It can be expanded!"

"Exactly," Sax ump ki's tone betrayed pride in his plan.

"Is there more you wish to tell me?" Cha wentz asked, pointedly.

"That is enough for now. The first thing is to take down the trees and haul the cut logs to the village. That will be hard work for all the men who are willing to help."

"How many do you think will come?"

"With the fishing season due to start, I hope for forty."

"Good," Cha wentz replied, "forty good men, warriors, strong and able, can accomplish much."

Just then the call sounded. "The ducks are ready."

As they had done each spring when the migrating ducks flew in to feed and rest, the people feasted. As far back as tribal memory could reach, the arrival of the ducks had been a time for celebrating. The ducks told them that the north wind had lost his hold over the land, and soon spring buds would swell and burst, greens could be gathered, and the land and sea would give of its bounty.

The villagers assembled and were handed the barbecued ducks, according to their position. Warriors ate first, then men of other skills, then women, and last the children. There was plenty for all.

Cha it zit waited his turn. His mouth watered as he watched the first eaters, including Sax ump ki and Cha wentz, tear at the crisp brown skin and then relish the flesh inside. They ate as men deprived, as if there would be no more, ever. When his time came, he did the same, ripping the meat with his strong, young teeth, and later tossing the bones to the dogs.

At last the orgy was over. The people lay on the ground, completely sated after a winter of dried, reconstituted foods, and the threat of starvation always hanging over them. Some groaned, others slept.

It was late the next day before the villagers awoke. No one really wanted to eat, but the women, out of habit, stirred the embers of their fires and rummaged through their nearly empty food boxes.

Cha wentz emerged from his part of the house and stretched. Cha it zit watched him from beneath his warm bearskin robe.

"I wonder what it is like to be married," he thought, "to have a woman to sleep with. I would ask Cha wentz but he would only laugh and say something I wouldn't want to hear."

Cha it zit had been thinking such thoughts for some time. It disturbed him; he tried to push them from his mind, but they kept coming back. He felt his face. That was another thing he had found himself doing. There was definitely a stubble there. The clamshell that Sax ump ki had told him to find lay in a storage box nearby.

"Maybe it is time to use it," he thought. Tentatively, he picked it up and closed the two valves over a facial hair. Then he pulled. "Ouch!" It hurt, really hurt. "I learned to ignore pain," he told himself. "That is one great thing I learned in my training." He pulled another hair. "Ouch! I must concentrate – my body does not feel pain." Again and again he pulled, until his face felt smooth as a child's, but it stung as if he had fallen into a nettle patch.

Preparations began for the gathering of food. Those women and girls who were to go to the islands for the annual camas harvest packed their digging sticks and mats into canoes. Later they would travel on to the clam beds. Those men who would fish made last-minute repairs to their nets, checked fish hooks, harpoons, spears, to see that all were ready.

It was up to Cha it zit and What la cum to set up a temporary camp and provide enough salmon and other sea creatures to feed the family for the year to come. Sax ump ki had said he would help, but Cha it zit knew where his great dream lay.

"He will build that wall even if he has to do it by himself," the boy thought. So he and his uncle loaded their canoes with the necessary equipment.

As was the custom when fishing or hunting, What la cum had not visited any of his wives for several days. He had bathed each morning in the coldest weather, as had Cha it zit, so that the spirits would find him acceptable.

When all was ready, they said their good-byes to Tseutsi and Sta nek and the younger children, and then climbed into their waiting canoes.

It was hard to make headway in the heavily loaded canoes. The nets and gear, packed carefully to balance the load, lay snugly around them. Where two should have been paddling, Cha it zit sat alone in his newly won canoe. His uncle, too, he noticed, was having difficulty.

When at last they reached the cut, Shalahsheen, where the water came and went, they found it dry. They beached their canoes and stretched out on the sand to rest.

"The water will return as the tide comes in," What la cum said. "Let us wait for it."

"That is wise," Cha it zit agreed, grateful for the chance to drop his paddle. "We will pass close to Temxwigqsan," Cha it zit remarked casually.

"That is true."

"I wonder how the wall is coming."

"It will have barely gotten started. There are many trees to cut."

"I thought that maybe we could see, anyway."

"We have fishing to do. They have a wall to build." What la cum's words were stern.

"True, uncle," Cha it zit soothed him. "We must not shirk our duty."

"I hear that old Xecusem, the headman, has a daughter," What la cum ventured.

"He might." Cha it zit tried to sound disinterested. "I have not seen her."

"But you would like to?" his uncle persisted.

"Maybe."

"The water is coming back," What la cum said, pointing toward a trickle of water which was working its way through the cut toward the bay. "Soon we can travel. Perhaps there would be time for a quick look at the project."

The trickle grew to a stream and the water in the depression deepened. The two pushed their canoes into it and shoved them, grating occasionally on stones, toward the open water of the strait. Then with the help of the incoming tide, they made good progress toward Temxwigqsan.

Smoke was rising from a stand of cedar trees at the rear of the village. Occasionally there was the sound of stone on wood as an axman cut through the char of fire. Then one could see a tree tremble and fall.

"It is sensible to burn a tree at the butt until it is weak and ready to fall," What la cum told Cha it zit. "Then a man with the spirit of the cedar can easily induce the tree to fall where he wants it to go."

But Cha it zit wasn't listening. He was watching the figure of a girl as she carried water toward the double longhouse. From the distance it was hard to see her features, but he noticed that she was sturdy, perhaps a bit fat. But that was to be expected when one remembered what her father looked like, he thought, visualizing the man's gigantic belly.

Could we get closer? He wondered. Yes, he could pretend to be looking over the area where the fort was to be built, but he must hurry, she was already nearing the longhouse. Quickly he thrust his canoe toward the shore, grounded it, and leaped out. Tired as he was, he raced toward the house; but she was nowhere to be seen.

Disappointed, he poked around the area, and then returned to the canoes where What la cum waited.

"Well, did you see her?" he teased.

Cha it zit frowned. "I saw the place where the wall will be. That is enough."

No one from the village seemed to notice their presence, so they angled across the strait toward the island, Samamao. It rose like a huge whale from the water. In the center, a hump, the dorsal fin, reached abruptly skyward. The sun was sinking low, and already it was blotted out by the hump.

"Samamao means high mountain," What la cum called out from his canoe. "It is well named. We must make Moltona by tonight and already the mountain is blotting out the sun. I should not have listened to you. We lost time at Temxwigqsan. It will be dark soon."

"Remember," Cha it zit called back, "when we round the point at Tcawogs, we will be on the side of the sun. It will be light again."

On they went, past beds of floating kelp where sea otters played. Cha it zit chuckled at their antics.

"Brothers," he called to them, "you make me laugh." Then one floated by on her back, holding a small pup on her belly. "Sister," he said gently, "you make me smile."

He felt that of all the creatures who shared the world, the sea otters were the most like him. "If I should become another kind of being," he told them, "I would be like you."

On the other side of Samamao, the sun was still shining, as Cha it zit had predicted. They glided past the point where the temporary fishing village called Tcawogs stood. He could make out the framework for mats that would be hung to give shelter. The water was phosphorescent. Each dip of his paddle released clouds of luminescent bubbles. Here the sea flowed like a river across reefs. It was clear and he could look deep into it. On they went. The sun dipped lower and then with a final salute, a flash of color, it dropped below the edge of the distant island.

"That island, between us and the sun, is the one our people came from," What la cum called. "Swelax is its name."

"I have been there," Cha it zit answered, "on my spirit quest." As it did so often, his mind snapped back to that time. Almost a complete circle of moons had passed since then, when he had set out alone in his small canoe to seek spirit power. How different his life was then. His whole life would depend on the success of his quest, and he had chosen to try for the most difficult spirit to find, Tiolbax. The agony, the endless diving, the hunger and thirst. He recalled the terrible pain of it. Now he had the power. He had proven it.

Then he thought of Tselique, as he had so often since his ordeal. The shaman, his great-uncle, had guided him through all of the challenges he had faced. Now his body lay alone on the island he chose to call home, in the place Cha it zit and Sta nek had laid him.

The *spirit boards* were there too, in the secret place where Cha it zit had hidden them. He conjured up a picture of them in his mind, misshapen paddles with designs of dots and a figure on them. Other items possessing powers Cha it zit had no knowledge of lay in the box Tselique had always stored them in. The boy had no wish to disturb the medicine man's belongings; even the thought of them lying there filled with power upset him.

"Some day you will be a shaman," Tselique had told him. "Then you will know what to do with these." Cha it zit rejected the thought and forced his mind to seek another direction.

Looking down into the water, he saw the shape of a fish moving slowly through a bed of seaweed. Suddenly he realized he was hungry. They hadn't eaten all day. He searched through his gear and found some cedar branch twine and a wooden hook and stone sinkers. He stabbed a piece of dried salmon on the hook and began to troll as he paddled. Before long he had caught several nice fat ling cod.

The sky became violet as they came to the bay. There were the drying racks and house frames that indicated a summer camp. "Moltona," What la cum announced. They pulled the two canoes as high on the beach as possible and tied them securely to beach logs. Then they began to set up their camp.

Cha it zit took a tightly wrapped bundle of inner cedar bark from the canoe and removed a large horse clamshell from it. Inside were coals which glowed as the air breathed on them. From these he built a fire. Before long, the fish were roasting on spits.

After the meal, What la cum took some dried *knik knick* from a pouch he wore slung about his neck and stuffed it into a soapstone pipe. He leaned back against a large beach rock and puffed contentedly.

"Uncle," Cha it zit began, talking in a disinterested sort of way, "I don't want to appear to be to inquisitive but I have been wondering about your new wife."

"What do you wish to know?"

"You seemed to get her so easily. You just went away for a while, and then you came back with her. My father told me that I cannot choose my own wife as you did. He said I must marry someone from a great family to give us prestige and wealth."

"What he says is true, so far as it goes," What la cum answered. "This wife is my third one. She is a small wife; she is not as important as a first wife, so she was easier to get. There were other concerns, too, things which have affected her life, so her people let her go without many gifts and much discussion."

"What other things, uncle?"

"That is not for me to say. What she wants known, she will tell."

"She must be from a wealthy family; she has a slave girl to attend her," Cha it zit observed.

"This much I will tell you," What la cum offered. "The girl is only part slave. My wife is her mother."

"How can that be?"

"She must tell you, if she wishes you to know."

Cha it zit thought about the conversation for a while. He was puzzled but didn't want to offend his uncle by prying. So they sat in silence.

The fire was burning down to coals and except for the immediate area, all was in gloom. Night had fallen, and one by one, stars appeared in the darkening purple sky. The lapping of the water disturbed the silence of the night. In the distance, high on the hump of Samamao, a wolf howled. The sound was answered.

"You are curious," What la cum said at last. "There is no harm in that. The man who does not concern himself with the affairs of another will be a friend to no one."

"Yes," the man-boy admitted. "Instead of telling me anything about your new wife and the slave girl, you have only created new questions. They are now of our family. What am I to think of them?"

"You are to judge them for themselves. Is that asking too much?"

"No," Cha it zit admitted, "but at least tell me how you got her for a wife so easily."

"It was not as easy as you think. The S'klallums are allied to us, but they are also different. Their customs do not always match ours. I will tell you this about her. She has strong powers, not as strong as a man can have, but for a woman she is to be feared and respected."

"I believe that," Cha it zit admitted, "when she greeted us, she stood tall and proud. Yes, I knew she was an unusual woman." Then he had an inspiration; the idea seemed to come from outside his mind. "Is she a medicine woman, a shaman?"

"You would discover that soon enough, so I will tell you. She has healing powers. She works not as a medicine man would but as a healer with special plants."

"That is good. We have no one to help us since – " Cha it zit stopped and sucked in the name about to be spoken.

"I know," What la cum said quickly. "It is dangerous to even think of that person. It could bring his ghost back to cause terrible things to happen."

"Yes," was all Cha it zit could say; words choked in his throat. Why couldn't he forget that person? Why did he always seem so near?"

Just then a raven cawed in the darkness.

"That is odd," What la cum remarked. "Why would a day bird such as a raven speak now?"

Cha it zit wondered, too. "Unless," he thought to himself, "the raven, the representative of Tselique's shaman power, was Tselique's ghost." He welcomed and shuddered at the thought.

The fire had died to a few embers and with the moon just rising over the hump of Samamao behind them, the restless waters rising and falling in gentle rollers became darkly visible.

Cha it zit, staring out toward the sea, could see the vague outline of the island of their origin. He was idly contemplating the tribal lore of the first people when all his senses sounded an alarm. A faint shadow passed only a short distance from the shore, then another, and yet another.

"Shush," he warned his uncle as he leaped up, kicking sand on the dying fire.

"What?" What la cum asked.

"Come," Cha it zit grabbed him and they scrambled for the dense brush above the beach.

They crouched behind a fallen log, half under it, and waited. There was no sound, only the night wind sighing in the branches of cedar trees. Half the night they waited there.

"If they were Yakultas, they have gone," What la cum whispered. "Are you sure of what you saw?"

"Yes, I am certain that I saw three canoes."

"They are reconnoitering," What la cum said. "They would not attack with so few warriors."

"That means there are more. They are planning something."

"It is the worst time for us. The warriors are building a fort, there is no one but a few old men and those who fish and hunt left to protect the women of the village. Even we, who are useless in battle, are not with our families. What can we do?" What la cum worried.

“We must think,” the boy said.

They sat in silence. Then an idea came to Cha it zit. “The Yakultas don’t want to invade us yet. They could have killed us and taken our canoes. Instead, they slipped past us. Maybe they are waiting for more to come. That will give us more time.”

“You could be right,” What la cum agreed.

“It is light enough for me to see, with the moon rising. You stay here, build up the fire so that they will know, if they are around, that we are still here. I’ll take my canoe and warn the people at Temxwigqsan. There are many warriors there, they can find the Yakultas before they can attack our villages.”

What la cum did not answer. He evaluated the idea carefully. Cha it zit waited in silence wondering why it took so long.

“You are young,” What la cum said at last. “It is a man’s place to go. I will do it.”

“Is it being any less a man to face danger here, exposing yourself so I can get away?”

“All right, go then,” What la cum agreed reluctantly.

Together they unloaded Cha it zit’s heavy canoe, working in the dark, quietly as possible. Then the boy pushed his craft out into deep water and slipped out into the darkness. He looked behind him and saw the fire leap into flame. The light was like a beacon in the darkness.

Cha it zit was tired. The trip to their fishing camp, Moltona, had been an arduous one. He had not slept for a day and a night. As he pulled the canoe, his arms felt numb, but he kept moving, steadily. He dared not sing the paddling song to help send the canoe over the water. Yakultas would surely hear his voice carried on the sea breeze.

The sky was beginning to lighten in the east when he reached Temxwigqsan. When he entered the larger of the twin houses he saw that warriors were sleeping everywhere.

In the gloom it was hard to miss stepping on the bodies lying on mats on the dirt floor.

"Sax ump ki," he called out, not knowing how to arouse the people.

A figure stirred and his father sat up.

"Eh," he said dazedly, "did someone call my name?"

"It is I, father; I called you. There are Yakultas nearby."

Sax ump ki jumped up, amazed to find Cha it zit there, alone, and questioned him. Cha it zit told him what he had seen.

"Are you sure?" Sax ump ki asked.

"I saw three canoe shapes, that is all I know."

Sax ump ki wondered, "Is it enough to go on? The word of a boy that he saw what could be canoes in the dark?" Then he acted.

"Wake up – wake up," he called out, picking up a paddle and banging it on a house post.

"My son comes with news." He nodded to Cha it zit.

"I saw three canoes in the dark last night at Moltona. They may be Yakultas. Our women and children are unguarded at the villages."

There was a hasty council of war. Sax ump ki was asked to make a plan.

"Yes," he said, "I have a plan. Ever since the massacre at the island where many of our people were slain and taken as slaves, I have been thinking of how to defeat the Yakultas."

"We will hear it," the leaders said.

"Let us first post watches so that we will know the whereabouts of those marauders. Let us place canoes in readiness so that when the sentinels spot the enemy, the canoes can be launched. Only those men who are skilled at sea battles should man the canoes. Then let us hide warriors on high ground. When they find the Yakultas, our warriors will engage them in a sea battle, but they will also lure them to the beach. Then they will seem to be retreating, running away into the woods. The Yakultas will follow; then the hidden warriors will kill them."

"Good," the warriors agreed.

"We do not know when they will come. We must patrol and watch. We must not let them attack any other village, only here."

The warriors scrambled to collect their weapons and prepare the trap. Sentinels with runners to carry messages were dispatched.

Sax ump ki called to Cha it zit. "Son, you did well."

"It was my duty," he answered, inwardly glowing at the approval of his father. Suddenly he was very tired. He found a secluded place in the forest behind the houses and fell asleep.

When he awoke, the sun was high and the air felt warm. He lay there on the mossy ground for a while trying to collect his thoughts. So much had happened in the past few days. Bees droned, birds sang. He felt strangely contented.

"This is a good place. I feel as though I belong here," he said to himself as he stretched. Then he thought of What la cum. Had he survived the night? Had the Yakultas seen the fire and decided to attack? "By covering my escape, What la cum may have given his life," Cha it zit realized. "I should go back, but now I can't, it might disrupt the plan."

His body felt stiff and sore from the hours he had paddled his heavy canoe the day before. To loosen his aching muscles, he began to stroll about the area. He could see that fallen trees lay strewn about behind the double building.

"The warriors have been working," he realized. "Soon the walls will go up. I wonder if the enemy has been spying on the village."

Inside there were no warriors to be seen. The family of Xecusem were going about their daily lives as if nothing was about to happen.

"Maybe they haven't been told," Cha it zit thought, "or maybe they are acting as decoys."

An old woman called to him, "Come and eat, we have plenty for a sturdy young man like you."

He gladly sat by her fire. "Young man, she called me, not boy," he realized. "Yes, she was right, I'm not a boy." He looked at his feet. They were much farther away than they had been a short time age. People who had seemed tall to him, he could now almost look in the eye. And his voice, he was never sure how it would sound when he spoke. Sometimes it was a squeak, and then again it would be as deep as a man's voice.

"Young man, that's what I am." He liked the sound of it.

The old woman handed him a bowl filled to the brim with a thick clam soup. It smelled delicious, and he ate with relish.

He looked about the house. Except for the old woman, the people paid him no attention. Then, suddenly, the girl was there. If she noticed him, she hid the fact well. Cha it zit was free to observe her, her appearance, her movements, her industry.

"She is not good to look at," he decided, "not like the Samish girl with her wide, soft brown eyes and sweet mouth, but there are other things in her favor." He tried to think what they were. "She looks strong, but someday she will be fat. She seems to be industrious, that is important." He knew that a hard-working wife could produce wealth for her husband. "She seems to be good-natured," he observed as she smiled and patted a child on the head.

Satisfied, he handed the bowl back to the old woman and ambled down toward the beach. All was quiet; too quiet.

Just then, a runner raced into the village. "They are coming through the channel," he said in a hoarse whisper.

“How many?” Sax ump ki popped up from behind a log.

The man held up ten fingers. “That many canoes, each with many warriors.”

“Send out the warriors,” Sax ump ki ordered.

Immediately, a dozen war canoes were manned and racing out into the straits. They were paddled furiously toward the enemy, barely visible in the distance. From the shore, Cha it zit could see the two armadas converging. Then he could hear, faintly, the war cries as each warrior tried to intimidate his adversaries.

Gradually, according to the plan, the battle moved closer to the shore. Now Cha it zit could see it raging back and forth. The fighters had moved to the off-center of their canoes, causing them to cant in the water, with the hull rising on one side to form a shield against the spears and arrows.

Occasionally a man would topple out, an arrow or a spear in his body. Neither side appeared to be winning; the Yakultas were formidable foes.

The battling forces neared the shore. Cha it zit ran for cover. He had no weapons, and anyway, that was warrior’s work. He could see the waiting land forces, weapons at hand, ready to take up the fight once the enemy had been lured ashore.

The first of the Salish canoes beached and its occupants, now feigning fear, ran up to high ground. The Yakultas were close behind them shouting and brandishing war clubs. Their faces were painted grotesquely. “They look like fiends,” Cha it zit thought. “Maybe they really are.”

Then the waiting land warriors fell upon the Yakultas, clubbing them as they ran past. The fleeing canoe fighters turned back to help finish the task.

Cha it zit watched in horror. It was bloody beyond his wildest imaginings. Men, strong, brave men, were screaming and dying. Some were lying in their own blood. Others were twitching with their last breath. Gone was the glory of battle. When it was over, the dead enemies lay strewn about the village compound. One man was held apart and spared. He watched as his fallen comrades were beheaded. He knew those heads would stand as trophies on the top of poles around the village.

The man, reduced to a trembling puppet, was brought before Sax ump ki. Cha it zit moved closer so he could hear what was said and see what was done.

"We despise you," Sax ump ki said with hatred in his voice. "You and your people have brought sadness and desolation upon us. I am allowing you to live so that you can return to your home to the north among the Yakulta people. You must tell them of this day, and of what we, the Nuh Lummi people, did to you. The same thing will happen each time you come to kill, to steal, and to kidnap. Remember my words." Then Sax ump ki pushed him. The man fell to the ground. "Go, I say, and come no more," Sax ump ki bellowed.

The man ran, stumbling, to a canoe and departed.

No one spoke. The warriors looked about them. Never had they fought such a battle. At first they appeared to be stunned. Then with ferocious shouts, they stomped and sang their songs among the fallen bodies.

When the work of taking heads had been accomplished, the warriors dug a great hole and pushed the remains of the Yakultas into it. Then they covered them with gravel and rocks.

"It is done!" Sax ump ki said to Cha it zit, who had joined him. "Our family has been avenged. My hatred is burned. It is no more."

Salmon

Cha it zit prepared to return to Samamao. He was concerned about the fate of What la cum. Had he been attacked and killed by the Yakultas? He feared to know the truth, but he knew he had to go. As he was readying his canoe, Sax ump ki approached. He laid his hand on Cha it zit's shoulder.

"That was a brave thing you did," he said. "If you hadn't given the alarm, the Yakultas could have killed many people and stolen even more."

"I only did what had to be done."

"And what of my brother?" Sax ump ki asked. "What part did he play in this?"

"He made himself a target so that I could get away. I go now to see what happened to him."

"I will come with you."

When father and son beached their canoes at the fishing camp, they looked for signs of a disturbance. All was as it had been when Cha it zit left, except that What la cum was not there. They called and then beat through the brush, but to no avail.

"His canoe is here. There is no sign of a struggle. I cannot understand it," Sax ump ki said, a distressed look on his face.

"He is hiding; he will come of his own will," Cha it zit said, hope in his voice.

"You could be right. Now that I am here, I might as well be useful. Let me help set up the canoes for the fish."

"I caught some bottom fish, but the salmon haven't come yet," Cha it zit told him.

"I know it is early, but our food is almost gone and we need fish now. If we prepare, the salmon tribe may come to see us."

Sax ump ki walked to a far part of the beach. Several large boulders lay a short distance apart near the water's edge. "These mark my fishing site. No one but me and my family can put nets here. Someday, if you become a good fisherman, it will be yours."

Cha it zit was pleased. It was a good thing to belong to a family with such possessions, he thought.

"We must secure the two canoes out over the reef," Sax ump ki continued. "There are anchor stones on the reef itself, which we can tie the canoe to. Here, I'll show you." He drew a diagram in the sand of four rocks forming the corners of a rectangle. "One canoe is anchored in this way," he said, adding a canoe with ropes running from its fore and aft tied to the stones.

"I see," Cha it zit said, "the anchors hold the canoe steady in one place."

"Now, when they are in place, we hang seaweed on the ropes to disguise them, to help direct fish between the boats," Sax ump ki continued.

Cha it zit nodded. "But then how do we get the fish? Spear them?"

Sax ump ki laughed. "Some ancestor, I don't know who, or perhaps it was Raven himself, showed us how to get the fish. We will tie a net, the one we mended a while ago, between the boats so that it lies on the bottom and extends up to the canoes like a weir. When we see from the canoes that there are fish in the net, we quickly raise one side of the net from one canoe and toss the fish out of it into the other canoe. That way we can get many salmon quickly without having to spear or hook them."

"That is clever," Cha it zit agreed. "As you said, let's place the boats so they will be ready when the *salmon people* come."

Under Sax ump ki's direction, Cha it zit dove into the icy water to tie ropes to the anchor rocks. These he secured to the canoes, his and What la cum's, through grooves provided for that purpose. It was cold, exhausting work, but it was exhilarating as well and he did it with enthusiasm. Working as a team with his father was a new experience for Cha it zit and he put out tremendous energy in his effort to please.

When the boats were securely anchored in their places, Sax ump ki called to Cha it zit, "Come, son, let us see if we can find a few fish for our evening meal. Maybe the aroma of them roasting will bring What la cum out of hiding, if he is still alive."

They opened fat, blue mussels and extracted the sweet, orange meat to be used as bait. Then using Sax ump ki's canoe, they paddled out over the reef. Cha it zit shared his fishing equipment, and father and son trolled slowly across the cove of Moltona.

"I've got one," Sax ump ki announced, pulling up a fat red snapper.

"He is a fine fish," Cha it zit said.

Sax ump ki spoke to the fish, "You have given of yourself to feed our hunger. I release you so your spirit may return to your people." Then he smote it smartly with Cha it zit's fish club and the gasping fish was still.

Before long the pair had enough fish for themselves and What la cum, too.

They started a small cooking fire, and set the fish to cook on green sticks placed at an angle over the coals. The smoke carried an invitation back into the forest, and soon What la cum emerged, looking groggy but quite whole.

"Yumm," he grunted, "I must have fallen asleep. What happened? Why are you here brother?"

"Because of you," Sax ump ki answered, a gruff edge to his voice.

"Me? Why?"

"We thought the Yakultas might have come back and killed you," Cha it zit told him.

"Hah! No chance of that. After I stirred up the fire to draw their attention, I got away into the woods. I crawled into a half rotten log and pulled the bark over me. I lay there, not moving, for a long time. Then I must have slept. The smell of that fish brought me to."

Cha it zit tested a piece of fish. It flaked easily, was temptingly tender and juicy. "It is done," he announced as he handed skewered fish to each man.

"Well," What la cum asked impatiently, "what happened?"

"Brother," Sax ump ki slapped him on the shoulder, "it was a battle that will be remembered by our people for generations. Yes," he laughed, "storytellers will be relating around the evening fires how the Lummi warriors defeated the Yakultas. They will speak of it for as long as there are people to tell it to."

Then Sax ump ki told what had happened, giving full credit to Cha it zit for sounding the warning.

"Now that they have felt our arrows and clubs do you think the Northerns will leave us alone?" What la cum asked.

Sax ump ki chewed the fish in silence. The question was a thought-provoking one. Would they stay away, or would they take the defeat as a challenge? Was there more bloodshed to come?

"I do not know," he replied at last. "But I do know that we will finish the protective wall and we will keep watch."

"Listen," Cha it zit said some time later. A sound, more of a vibration, growing louder and louder, seemed to engulf them. It came at them from the quiet water, from the trees and from above, traveling down the mountain. The pulsating was like the beat of a giant heart.

"Drums," Sax ump ki said. "They are celebrating the victory at the village."

"It will be a wild night there. Those warriors will really put on a show," What la cum said.

"Show?" Sax ump ki sounded surprised at the use of the word. "Show? It will be no show. It will be real. Some of those warriors are members of a secret society. When they go into their spirit dances, terrible things can happen. I know."

"Are you a member?" Cha it zit was surprised that he dared ask the question. Those things were not spoken of.

"I was once," Sax ump ki admitted, "but since I have lost the power of youth, I no longer go to their dances."

The drums continued to reverberate through the night air. Cha it zit tried to imagine what was happening in the longhouse at Temxwigqsan. It was so far away, how could the drums be heard? "There is life in a drum, it can make itself heard," a voice inside him seemed to say.

Cha it zit, sitting between his father and his uncle, comfortably full of fish, gazing at the flames of the fire, felt a glow of contentment. Never before had his life seemed so good. He was now accepted as a man. His opinion was respected. He was respected. It seemed that all of the trials and hurts of the past were behind him. Yes, to be a man was a good thing, he thought.

"Tomorrow we try the waters for fish," What la cum interrupted Cha it zit's reverie. "Will you fish too, brother?"

"No, I will return to Swetquem to help bring the women. When the fish come, they must be here ready to process them. And I have a feeling that the salmon will come soon. You and my son will fish."

"What is it like when the salmon come?"

Cha it zit wanted to know. He had heard stories of great migrations where the fish gleamed silver in the water and the sound of them coming could be heard before they were seen. He wondered if such a thing could be true.

"You will see for yourself," What la cum answered. "It is better that way."

"Yes," Cha it zit said. "Why do they come?"

"It began a long time ago," Sax ump ki replied, "during the time before the *Changer* came to change all things to their present form. In those days supernatural beings roamed the earth, beings that could do wondrous things. There was a man who went out to try to find something to eat. He was hungry, for there was not much food in the world. As he wandered about, he came to a bank of fog. It encircled him so that he couldn't see where he was. Suddenly he saw a woman standing in front of him. She was very comely and she smiled at him. He decided to make her his wife. When the fog lifted, she went with him. He noticed that she wore an unusual hat which she never removed from her head. They arrived at the man's house and he told her he was hungry and could not find food. Upon hearing that, she removed her hat, which he found to be filled with water, and lifted a fat fish out of it. This she cooked for his dinner. After that, whenever he was hungry, she would take a fish from her hat.

"The man then grew greedy. He decided that his wife's talent would make him rich, so he demanded that she produce many fish so he could trade them for other goods.

"For a while she agreed. Though she grew tired, the man wanted more and more. One day she took her hat and disappeared into a fog. But as she was leaving she told him that once a year she would toss her hat into the sea and all the fish would come."

Sax ump ki stirred the fire, then said, "But that was long ago. Things like that don't happen any more."

"We know that there are five tribes of the *salmon people,*" What la cum said. "Each of them is different, and they come at different times."

"Where do they come from?" Cha it zit wanted to know.

"Each tribe has its own village under the sea. They are like us most of the time. They live as we do and have a headman to lead them. When it is time to come to us, they change their appearance as we do when we put on a blanket," What la cum replied.

"How do they know when to leave their villages?"

"They send out scouts," Sax ump ki said. "That is what we are waiting for now. If the scouts are well received, they will tell the rest to come. It is very important that the first fish caught be treated with great respect. Otherwise the rest of the fish will not come to give themselves to us and we will starve."

"What do we do for the first fish?" Cha it zit asked.

"You will see," Sax ump ki told him. "It is the headman who must be responsible for that. I will return in time. If I do not, my brother will act for me. He will tell you what to do then. Now it is time for us to rest. There will be much for each of us to do tomorrow."

The three wrapped themselves in their blankets and lay on the sand. The drums continued their hypnotic beat, beat, beat, far into the night. They seemed to have found their way into Cha it zit's head and were pounding against his skull. When at last he fell asleep, he dreamed that he was in a longhouse, lying on the floor while dancers leaped and stomped on his body.

Sax ump ki left the next morning. Cha it zit watched until his canoe rounded a point and was lost to sight.

"We will take turns," What la cum said. "I will sit in one of the canoes until the sun is overhead, watching for the first salmon. Then you can have a turn. We'll use a hook and bait; no point in using the net until the salmon tribe comes."

"Which tribe will come first?" Cha it zit wanted to know.

"The best one, the finest eating fish of all, the Chinook."

For two days they took turns angling and watching for the first salmon to appear, but all they caught were a few rockfish.

On the third day, the Swetquem people arrived with Sax ump ki, Wanana and Latsi in his canoe, leading the others.

Immediately the fishing camp came alive with activity. While the men helped to unload the canoes, the women began to set up the temporary shelters. They tied mats to the frameworks already in place, or erected new ones, as the need arose. Then they stored their digging sticks, mussel-shell knives, baskets, and water buckets inside. The instant village ranged along the beach, each shelter with a view of the water and a bit of ground around it.

That done, the women inspected the drying racks that stood on high ground where the sea breeze could hasten the process of preserving the salmon. They made repairs where it was necessary and then turned themselves into beasts of burden. Wood, much wood, was needed to keep the smoke fires going day and night until enough salmon was hard-smoked to last through the winter season, a time when great quantities of food would be needed to feed guests as well as the family.

It was almost nightfall by the time the weary women had finished their preliminary work. Tired as they were, they now had to think of preparing the evening meals for their hungry men and older children. Soon small fires burned along the pebble beach. Bottom fish – cod, flounder, sole, even halibut – which the men had caught by trolling from the canoes were cleaned and placed to roast.

The men, meantime, had been securing bait for the fishing next day, idly chatting about their prospects. They took mussels from the rocks, prying them loose from the strong filaments that held them tight; they dug succulent steamer clams from just under the gravel bars; dipped small fish from tidepools. Then they gathered on the beach, hunkered over their catches, and prepared the bait.

"We'll troll for the first of the salmon," they agreed, "he will come any time now – a big fat Chinook, a noble of the tribe, maybe even the headman himself will come."

Cha it zit watched the activity with interest; that is, when he was given the time to watch at all. The women, not yet granting him the status of a man, asked him to bring water, help with the largest logs for the fires, and to run errands. The men, not yet accepting him as a fisherman, asked many services of him, too.

"I thought that I had become a man," he complained to himself. "I thought I had earned that station – now I am being treated as a boy, 'Run here, run there, do this, do that!' When will I be seen as a man?"

He sat down on one of the rocks which designated Sax ump ki's reef-net location, and looked dejectedly out to sea. The large island, Swelax, the one from which his people had come, rose from the water. Half shrouded in sea fog, it appeared as a purple vision. It was there, he recalled, that he had suffered and dived for his spirit vision. That agonizing experience, as much of it as he remembered, gripped his mind. It had often done so since the day he had found the power he sought off the shore of the island, Samamao, on which he now sat.

"It doesn't seem to matter," he thought, "that I have power, they still treat me as a child."

Sax ump ki left the group of men, who by now were idly discussing the weather and filling their pipes for a smoke, and approached Cha it zit.

"Son," he said, "you are troubled."

"Everyone treats me as a child," he said, speaking before he thought.

"You are treated as a child if you act as a child," Sax ump ki answered.

"I don't understand it. A few days ago, I felt I was finally a man. I was treated with respect, I was even allowed to give my opinion. What has changed?"

"The circumstances. Have you been on a food-gathering trip before?"

"This is the first. Last year I was on my spirit quest; before that I stayed behind the help the old ones."

"Then you are a child in this situation. You must learn and prove yourself. If you try hard, you may find a fishing spirit to add to that which you already have."

The evening meal was ready. Wanana and the other wives of Sax ump ki had prepared the fish in such a way that it was succulent, still dripping juices. They had gathered wild strawberry leaves and steeped them in boiling water to make a sweet, refreshing tea. When Cha it zit and Sax ump ki had seated themselves near the fire, Latsi handed them portions of the fish, wrapped in leaves of the big-leafed maple tree.

The next day the men and older boys left to go their own ways, trolling for salmon, but on the lookout for a seal to harpoon, or halibut to bring up from the depths. The women, always busy, took their baskets to search along the fringes of the heavy brush for early berries. The salmon berries were ripening, and small wild blackberries could be ready in sunny patches.

In the evenings, they ate whatever they had gathered or caught and relaxed by their fires. Sax ump ki worried. Some years the salmon didn't come. He knew that was because they hadn't been treated well the year before, and he also knew that a poor run of fish meant disaster for the people. "This year," he thought, "the salmon may not come; they are late, now." He kept counsel with himself, but What la cum watched him intently.

One day he said to Cha it zit, "Your father is concerned. He fears the fish will not come."

"I know," Cha it zit answered, "what can we do?"

"If we had someone here with power over the fish, maybe he could bring them."

"Is there no one like that here?"

"No one with that kind of power."

Cha it zit thought for a while. An idea came. "You said your new wife is a medicine woman. Could she help?"

"She is a healer, not a shaman."

"Ask her," Cha it zit insisted. "She may have more power than she admits to." What la cum's face hardened and so Cha it zit said no more.

Each day the captains of the reef-net canoes dressed themselves in the traditional costume worn to greet the first of the great salmon run. Clothed in strands of goat wool and duck down, they watched from their canoes, waiting for signs that salmon had entered the submerged nets. The wait seemed to be endless and there was the unvoiced worry that the fish would not come at all.

Then suddenly they were there. "Watch them. Watch them," an excited captain called out. "Lift, lift," he ordered his crew. The water churned as the net was raised and a rainbow of spray filled the air.

On the beach the women dropped their work with shouts of joy. They had long been prepared for this moment, this time of welcoming the first of the salmon tribe to enter the nets. They knew that if all was done properly to honor the fish, a multitude of them would follow.

Winter ferns had been gathered, soft duck down filled storage baskets, cedar bark had been shredded, and kernels from a special plant had been collected to be ready for the important *first salmon* ceremony. Now they called to the children who were to have an important part in the ritual.

"Quickly, come," their voices sounded along the beach and across the water. As the children arrived, breathless and excited, their backs were painted red and duck down was strewn in their hair.

When thc canoe bearing the first salmon of the year reached the shore the people were already assembled there.

As it was his right to do, Sax ump ki called out the first greeting, walking into the water even before the canoe was beached. "We welcome you. We greet you as an honored guest. Our people are happy you have come and have made ready to receive you. Our women will feed you and honor you. Our elders will sing spirit songs for you. Our children will treat you with respect."

Then each salmon was placed on the outstretched arms of a child who steadied it by clenching its back fin between his teeth as he had been instructed to do. Then, according to custom, the bearers of the salmon side-stepped up the beach to where the women received the fish and lay them gently on pallets of winter fern fronds, their heads toward the water.

Then the oldest among the people placed kernels of the spice plant, duck down, and a small stone bowl of red paint before the salmon. "Now we feed you," they said to the fish, "that later you may feed us." Then cedar bark was set on fire and the food was burned so that it would accompany the fish to the *spirit village* of the salmon, their home under the sea.

The women, with arms and faces painted red and duck down in their hair, wielding new sharp mussel-shell knives, slit the fish along their backbones and pulled each one open, carefully cutting the flesh away from the skeleton. The bones, intact, were returned to the fern pallets.

The men began to dig a trench in the beach sand and to build a fire the length of it. While the wood was burning to coals, an elder, who was versed in the ritual songs, sang the salmon spirit song to honor the fish. It was a long song, ancient beyond memory. At the end of each strain, he named a place where salmon were habitually caught and put a gift for the fish of that place into the water.

Cha it zit, who was among those tending the fire, became completely immersed in the solemnity and importance of the ceremony. His mother and sister, along with Xedewa and her daughter, Tsexad, working with the other women of their village, seemed like strangers with their red-painted faces and arms, performing a familiar task with an unfamiliar meaning.

The old man who sang to the fish was from another house. Cha it zit thought of Sta nek. Before he became so old he would have been here and he would have been the one to sing. And before him it would have been Tselique. "Now we have no one in our house to do it," he realized. Soon Sax ump ki will be an elder, then the task would fall to him. Then what? "Will it be mine someday? Will it be me carrying on the traditions of our people in my person?" Somehow, it was unthinkable. He was not worthy. He stirred the embers and they exploded in a shower of fiery sparks.

The people waited quietly. The old man sang endlessly. All must be done properly, according to the custom, for much depended on the fish being pleased.

When the fire had burned down into coals, those responsible for cooking the salmon placed green ironwood sticks across the trench and with great care placed the fish upon them. Soon the aroma of barbecuing salmon filled the air, but no one ventured to touch or taste the fish until those whose privilege it was removed them from the branches and placed them on fresh beds of ferns in rows on both sides of the fire.

"Come, children, eat," the women called to all of the children. As they were eating, kneeling before the salmon, old men passed behind the children, painting again their backs to purify them and make them pleasing to the salmon. After that, all the people partook of the first salmon feast.

The charred sticks on which the fish were cooked, together with the bones, were gathered up and carried to the water. There, Sax ump ki said to them, "Return to your village under the sea. Tell the salmon people how we have treated you this day. Tell them that we want them all to come to us. We will treat them well. If they will give themselves to us for food, which we need, we will do for them as we have done for you; return their bones to the sea. In your villages they will become whole again, as will you."

Sax ump ki waded out into the water and deposited the bones and wood on the waves. "Go," he called, as they vanished beneath the water, "but return quickly."

"It will be several days before the fish come," Sax ump ki's first wife said to the other women of his family. "Let's take our digging sticks and test the camas beds. When the salmon run, we can do nothing but prepare the fish."

The camas beds were on several small islands quite visible from Moltona. They rose from the inland sea like scoops of debris left over from the material used in building the larger islands of Samamao and the island of origin. Grass grew on their low, open surfaces and among the stems, blue camas bells waved in the breeze. Lovely as they were, the blossoms did not interest the women; deep under the sod, each plant possessed a bulb that was sweet and nourishing.

Those bulbs were so important to the people that the possession of even a few square feet of a camas bed was a prize that was passed on from generation to generation. As was true of all the natural resources available to the people, it was the privilege of taking those for use, not the land or sea that produced them, that could be owned. This the people understood and accepted.

Soon a flotilla of small women's canoes could be seen heading for the camas beds. As they paddled, the women sang. This was to be an outing, a chance to get away from the routine of drudgery that was their daily lot. True, there was no harder work than digging camas; the ground had to be pried up, bit by bit, the sod laid back to expose the roots from underneath. Even then, only the mature bulbs were taken; the bulblets were left intact to grow another year. But this day was to be different; they would dig some, but mainly, they would see if the bulbs were ready.

The women scampered ashore, chattering and laughing at their private jokes. Wanana came more slowly. Latsi waited for her. "Are you well?" she asked, worried.

"I am all right," her mother answered, her breathing labored. "Go along with the others, I'll just stop for a while. Go now!"

Latsi hesitated, then ran up toward the open meadow where the camas grew. Some of the women were already digging up the sod with their sharpened, fire-hardened sticks.

"Where is your mother?" Xedewa asked her.

"Back by the beach; she needs to rest."

The medicine woman left the others and ran down to where Wanana was sitting on a log.

"Is the baby soon to be born?" she asked.

"I am not sure," Wanana answered, her face tinged with pain. "I have discomfort." She clutched at her abdomen.

"There is a seaweed that will help. I'll look for it – it grows at my village; maybe there is some here too." Xedewa hurried to the water's edge.

Wanana stifled a groan. "Don't let it be now," she moaned.

Xedewa returned with a handful of rubbery green seaweed, dripping sea water. "Here, chew this, you will feel better."

Wanana took the medicine and chewed with vigor.

"Do you feel the pain easing?" Xedewa asked anxiously.

"It is better," Wanana tried to smile. "You are kind. I am happy that you came to us."

Xedewa put her hand over Wanana's. "Those are the first words of friendship I have heard since I came. Thank you."

Xedewa felt Wanana's abdomen, probing and pressing gently.

"I do not think your baby is ready to come," she said, "but you must be careful or it will decide to be born too soon. When we return to Moltona we will build a birth hut for you in case you will need it. There are things that we must do to be ready.

"That is wise," Wanana agreed.

"Have you been observing the proper taboos?"

"I bathe every morning. I have abstained from red meat and fish, I have not been with my husband since I knew there was to be a child," Wanana answered.

"And your husband?"

"I do not know. Since the beginning of the wall building, I have spoken to him very little. His mind has been on only that one thing."

"You do know how he eats."

"When he eats at my fire, he consumes what he wishes, meat, fish, it makes no difference," Wanana admitted.

"That could be the problem," Xedewa observed. "He may have angered a spirit. I will ask my husband to speak to him – that is acceptable between brothers."

"I wish you would not –" Wanana began.

"Hush," the medicine woman said sternly. "Do you want to bear a live, healthy child?"

"Of course, that is every woman's wish."

"Then it will be as I said," Xedewa said firmly.

Wanana rose and the two walked together slowly up the incline on a path that had been worn through the ages by the plodding of the hard-working women.

The diggers had scattered, each to her own patch of camas ground. Wanana and Xedewa found Latsi and Tsexad, Xedewa's half-slave daughter, digging together, laughing at each other's dialect as they worked. Neither of the girls were experienced and the work was going badly for them. A small heap of bulbs lay on a mat between then. Some were small, some were mashed, only a few were mature, full-sized tubers.

Wanana smiled indulgently. "They are willing, they will learn."

As Xedewa and Wanana knelt nearby, Wanana explained the ownership of the bed.

"It really is Tseutsi's. She is no longer able to dig, so we, the wives of her sons, dig for her. What we get we share with her so Tseutsi and Sta nek have camas without having to work for it."

Xedewa sat back on her heels and looked about her, observing the other wives of What la cum and those of Sax ump ki as they plied their digging sticks, turning the sod and extracting plump bulbs.

"Tseutsi has seven daughters-in-law. She is very old. How will she pass on her ownership? This camas bed is valuable. The soil is deep and good. The bulbs are plentiful here. What is to become of it? Does she have daughters of her own to pass it on to?" Xedewa asked.

"She had several," Wanana answered. "One died in childbirth and the others, they tell me, were taken when the terrible sickness came. I suppose that Klatok, as the first wife of her oldest son, my husband Sax ump ki, will be the one to receive it."

They glanced across the field to where Klatok was digging at a steady pace, concentrating deeply on her work. She was a heavy woman and she stabbed the ground with her stick in a manner that suggested that even the bulbs under the sod were subject to her will.

"You don't care for her, do you?" Xedewa asked.

"She is what she is," Wanana answered after giving the question some thought. "She is Sax ump ki's first wife. She is high born and came for a great bride price. Many gifts were exchanged when the marriage was agreed upon. Her status among us demands that we respect her."

Xedewa laughed, "But you haven't answered my question."

"The answer is of no consequence," Wanana's tone ended the conversation.

The camas was ready, the diggers had learned, so what was to have been merely a time of testing became a serious work party.

All day the women and girls hunkered over the backbreaking work. As the sun sank lower, the piles of sweet bulbs grew higher.

At last Klatok painfully straightened her back and stood erect. She scrutinized the heaps of camas and then noted the position of the sun. Already it was casting long shadows from the stand of trees above and beyond the beds.

"It is time we left," she called out in her imperious way. The others stopped and stretched their tired muscles. Voices answered from other parts of the small island. So, gathering the bulbs into baskets and rolling up their mats, the women returned to their canoes.

"You come with me," Xedewa said to Wanana. "My daughter Tsexad can go with Latsi in your canoe. I think the strain of paddling is bad for you. I am strong; you can rest on the way back."

Wanana was glad to obey this strange new woman, this woman with special powers.

The day had gone well at Moltona. Dusk was overtaking daylight as the women returned. They saw that the reef-net boats were all in place. Two by two, they lay anchored offshore in a row paralleling the beach. The men, including Cha it zit, were resting in the shade of a large maple tree. Some of the younger village boys, who had been brought along to learn the fishing techniques of the people, were playing a game of shinny on the beach. The onlookers cheered one player or another as he hit the ball with his stick toward his team's goal line, one of two lines visible on the sand. It was a rough game the boys were playing, and occasionally they whacked each other rather than the knot of hardwood they were aiming at.

The tired women beached their canoes and proceeded to carry their baskets of camas to a spot of high ground above the beach. The men rose to inspect their wives' day's work.

"It looks as if we will have a feast," What la cum observed, glancing at Klatok who stood over the harvest, hands on hips surveying the loaded baskets.

"Yes," she said, authoritatively, "we will feast, and then we will work."

"When we finished setting the boats, we took the small canoes and trolled for halibut," What la cum told the women. "There are several fish under the ferns." He pointed to a heap of freshly cut sword-fern fronds.

The women prepared the halibut, cutting them into chunks to roast on spits. Later they would cut the remainder of the large, flat fish into strips to dry in the sun and wind.

When the villagers had eaten, Wanana wandered off to find some salmonberries to appease her hunger.

Soon, all was quiet and the people slept.

In the early morning, the preparations for the camas feast began. The men dug a large trench on the beach. They started small fires down its length, which gradually they built into one continuous bed of coals.

While the women washed the bulbs, the girls were sent to gather bundles of seaweed and to cut armloads of sword ferns.

Then the men scooped the live coals from the pit, leaving red-hot rocks exposed to view. The girls flung armloads of the seaweed onto the rocks, and it exploded in a geyser of scalding steam. The ferns were placed on the seaweed and the camas was laid on top. More ferns were laid over it all, and then more seaweed. At last, mats were laid over it all.

Now the people had to wait. It took hours to steam the succulent bulbs to perfection. Not much work was done that day. The young people played games and the adults sat and chatted.

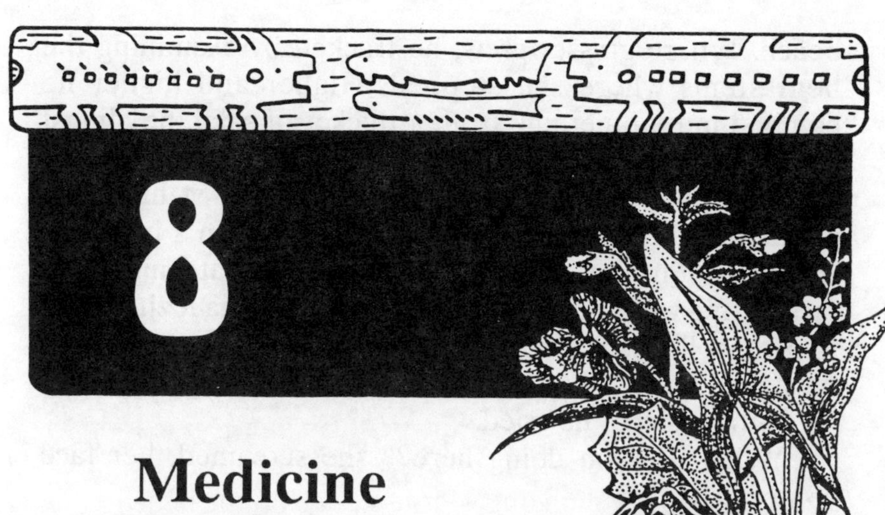

8

Medicine Woman

Cha it zit felt he was too old to play with the younger children on the beach, and he was too young to be fully accepted by the men. He would like to spend some of the leisure the day offered with Latsi, as he did in a time that seemed so long ago, but that was now frowned upon.

He wandered about the campsites, where the people sat in little knots, some of the women playing dice games, the older men smoking and telling stories about their fishing exploits, the younger ones listening. He saw a figure, small in the distance, moving toward a high meadow on the slopes of Samamoa's hump. It was a woman, and she carried a basket, held by a trump line from her forehead.

"It looks like Xedewa," he said to himself. "She must be going for healing plants. I wonder what she uses when she cures people?"

Without making a conscious decision, he followed her. It was easy to track her footprints in the soft soil above the beach. Where grasses grew, he tracked by examining the bent stems where she stepped. Higher and higher he climbed until, emerging from a thicket of salal, he almost stumbled over her as she stooped to dig out a root.

She jumped up and whirled, striking out at him. The strength of her blow sent Cha it zit reeling. In a moment, she was upon him, flailing with her fists, scratching at his face, biting viciously with her strong teeth. Cha it zit sought to protect himself, curling up in a ball. The blows rained on his naked back.

"Stop – stop," he cried.

"What are you doing here?" she screamed, her face contorted.

"I wanted to learn about healing plants," Cha it zit flung back at her. "Why did you attack me?"

Xedewa stared at him, her eyes shooting fire that seemed to pierce the boy's flesh. Then, gradually, the fire died and she became calmer. Her chest heaved as she sank slowly to the ground. Then she put her hands over her face and she cried. Great sobs racked her body. She rolled on her haunches from side to side in seeming agony.

All Cha it zit could do was watch, hoping the storm would pass. And it did.

"I need to explain to you why I acted as I did," she said at last in a stifled voice. "I have had great troubles in my life; they have left their mark on me. It is not good when someone startles me. Do not do that again."

"I won't," he promised, his tone emphasizing the fact.

"Come," she said. "We will sit on that rock overlooking the sea. I have been wanting to talk to you about many things. Perhaps the meeting here was not accidental, eh? The spirits are moving us toward some goal. I feel it."

"You are a strange woman." Cha it zit found her foreign speech hard to understand, but she, herself, puzzled him. There was no humility, no hint of servitude in her.

"I come from different people," she said, "and I am a medicine woman . . . no, I am a shaman! I do not bow to anyone."

"What was your trouble?" Cha it zit knew such a question was rude, but his curiosity overruled manners. "What la cum told me that you had some evil come to you, but that only you could tell of it."

"He spoke truth. I don't know why, but I feel compelled to tell you."

Cha it zit sat, looking out toward the mountainous island, the original home of the Nuh Lummi people. He waited to hear Xedewa's story.

Xedewa looked directly at Cha it zit and stood up, horror in her face. "Look at what I have done to you!"

He hurt, he knew, but he had learned to tolerate pain. He felt his face, and there was blood on his fingers when he touched it. His arms, too, were reddening where she had bit him. Tooth marks showed.

"Wait here," Xedewa ordered, as she ran off toward the shadowy forest.

"What a strange woman," Cha it zit mused. "I could have fought back. Why didn't I? Why did I let her pounce on me like that? What is her power?"

As he waited, he thought of the women of his family. They would never dare to raise a hand against a man, not a husband, surely, or a brother, or even an almost-grown son like himself. A man could beat a woman, that was expected, but a woman a man? Never. He was sure that Cha wentz hit his new wife; there were bruises on her back and arms that said so, and no one spoke of them. Except on their marriage night, when she scratched him as women did to show their purity, she had not raised even her voice to him. How could this unusual S'klallum woman be so different?"

Xedewa returned with a handful of leaves. They were thick and spongy looking with a pattern on them, yellow on green.

"We call this plant medicine," she said as she pulled the top layer of the leaves off. The insides lay exposed, juicy and tender. These leaf halves she placed over his wounds. "Now they will heal quickly," she told him. The juices soothed his pain immediately.

"Now, I will tell you of myself," Xedewa sat beside him again. "I was born into a noble family, the highest among my people. We had many slaves, and much wealth. By the time I had reached womanhood, my family had given three potlatches already. My confinement, when the sign that I had reached the time of childbearing came, was long and hard. When it was over, it became known that I was available as a wife. Many suitors came because of the prestige of my family. One was chosen. He was much older than I was, and was equally as powerful as my father. He too, possessed wealth. They both felt that the alliance would benefit themselves. The bride price, the gifts which that man paid, was very great, but the gifts that my family gave were of great value too.

"I was taken to live in his house. It was sad for me, and I became homesick. But worse than that, he was cruel to me. I grew to hate him. The more he abused me, the more I hated him. I thought of running away, but that would bring disgrace to my people." She paused.

"What did you do?" Cha it zit asked to encourage her to continue for the narrative was fascinating to him.

She did not answer for a while; and then she spoke. "I found a friend – more than a friend. My husband had many slaves. Among them was a young northern warrior who had been captured in a battle between the Yakultas and the S'klallums. He was pleasant to look at, powerful, taller than our people, with an air of arrogance about him; even as a slave, he carried himself with authority." Her strong features softened as she remembered him and she smiled, gazing across the shining water.

"He helped you?" Cha it zit prompted her.

"He encouraged me, told me things would get better. Strong feelings grew between us. As much as I hated my husband, that much I cared for this man. We plotted a way to escape, he from slavery, me from bestiality. One day, on the pretense of my going to bathe, and of him bringing in wood, we met and slipped away to where I had hidden a canoe. It was wonderful." She threw up her arms and laughed. "We were free. We found a place where a stream emptied into the straits and there we stayed."

"How did just the two of you survive?" Cha it zit couldn't imagine life outside the clan, the all-protective, self-sufficient clan.

"We started from nothing, but together we built a shelter; I made baskets, he made boxes, he fished and hunted, I gathered foods, clams, berries. We were happy. But it didn't last."

"Why? What?"

"My husband and his brothers found us. He had sworn that he would have his revenge. They came upon us suddenly, as you did me. They killed my lover and then they mutilated him. When my husband turned toward me, I ran. I ran as I never had before, and I escaped by hiding under a rotten log. When they finally left, I took the canoe and returned to my family home. By then I was pregnant with the child of a Yakulta."

"What did your parents do?"

"I had disgraced them. They had been told of how I had run away. My husband demanded his gifts be returned, and they were. Then he demanded more as damages. My people had given everything that they could before, but they managed to get more goods together, and to save the family position, they gave that too. They did not welcome me back, but they let me stay.

"When my daughter was born, my own mother stopped me from flattening her head, as all high-caste mothers do with their babies. 'No' she said to me, 'this child is half slave. She will grow up as a slave. She will live as a slave. She will have a round head to show her position.' I argued with her; I told her that the father had been born to a high caste, but she would not listen. I suffered more in my family home than I had with my husband. My people were poor because of me. Their prestige was gone because of me. I took my daughter and went into the woods each day. I wanted us to die, I think. It was while I was there that my first vision came."

"You had a vision?" Cha it zit was puzzled.

"It was a true vision. I received my first instructions. I knew I was to be a shaman, to heal others. But I knew I first had to heal myself. The rest is not for you to know."

She sat quietly. But Cha it zit wasn't satisfied; he wanted to hear more.

"That was a long time ago. Your daughter is now almost a woman," he ventured.

"Yes, those years were spent learning how to heal with the plants that grow. The vision came many times, and each time I grew stronger in power. People began to recognize my strength and came to me when a sickness came upon them. I began to pay my family back, in a small way, for what they had lost through me."

"How did What la cum find you?"

"My people heard that he had come looking for a suitable wife to take back with him. He had only a small bride payment, so no one was much interested, and besides that, he already had two wives, so the third would have little standing. But when they told me of this man I saw it as an opportunity. He and my father discussed such a marriage and they agreed that few gifts were needed. My daughter could come as my slave and no one need know who she is. I made only one request, that two of my nephews come with us to see how this new husband would treat me and to defend me if necessary. I vowed never to be so abused again, not as my first husband had done."

"And are you pleased with us?"

She hesitated before answering, "Yes, I hope you are pleased with me."

"We must learn to understand you," he said, honestly. "Then we will accept you fully."

"There is more I would say to you," she spoke seriously, "but the time is not right – it must wait."

He wondered what she meant, but kept his peace.

"Now, you must go. It is not good for us to be found together. I will finish gathering my medicines; they are for your mother."

Toward evening Klatok opened the camas pit. A wonderfully sweet aroma filled the air. She speared a bulb on the point of her digging stick and tested it by punching at it. It was soft. Then she popped it into her mouth and rolled her small, puffy eyes with satisfaction.

"The camas is ready," she howled. The sound carried up and down the beach and was picked up by other voices: "The camas is ready." Mats were laid out and the bulbs, now reduced to pulp, were heaped on them. As was customary at all feasts, the men ate first, then the women and last the youngsters. There was plenty for all.

The people ate with deep pleasure. This first taste of fresh baked camas was a once-a-year treat, one relished and remembered. They ate until they could eat no more. Then they rolled on the ground to ease their aching stomachs and finally slept.

The next day the salmon came. Those who had kept watch on the reef-net canoes called out: "They are here!"

The people ran to the beach. From the straits, beyond the anchored canoes, there came a sound, a swishing like the rushing of the wind. Then the sea glinted with silver and the water seemed to boil with the running of the salmon.

"The run of fish is great." They rejoiced as they hurried to take their places for the work they must now perform. Each had his place, his part in the vital effort to harvest the fish. The men best suited to spot the fish and those to work the nets, took their places in the reef-net canoes. The women, who would receive the fish and prepare them for drying, came bringing mats and knives. Those who would hang and smoke-dry the salmon tested their racks and lay their fires.

The great run, fish piling upon fish as they migrated toward the large river to the north, meant life itself to the people. It meant prestige and wealth, too. There would be no rest while the fish were coming through this route which they followed; that they had followed for as long as the people could remember.

The drying and smoking fires burned day and night as the women added fresh fish and took those that were dried down from the rocks. All day long, too, the men worked on the boats, their calls of "fish in the nets" sounding out along with the answering calls of the net tenders could be heard as they heaved the heavy nets up and over to dump the silver fish into their canoes. The piles of hard-dried salmon rose, and they were then placed in storage baskets.

Cha it zit had never worked as hard as he did during those days. Working with What la cum, it was his job to spot the fish and then help direct the net full of them into his canoe. "Will this never end?" he kept asking himself.

One day it did. All the boxes and baskets were full. A few late stragglers among the salmon tribe still swam past, but no one wanted to bother with them. The people were tired.

"We have done well," Sax ump ki told his family. "It is time to take the fish back to our house. After that there will be more camas and clams to be dug, seals to take, deer to

weir
Pink
Chinook
Chum
Sockeye
Coho
Steelhead
fern
camas
root digging tools
shellfish tools
salal
huckleberry
moon snail
limpet
mussel
salmon-berry
black-berry
clam
oyster
strawberry

bring down, all on our island of origin. As for me, I will inspect the stockade."

The people prepared to leave for home. Reef-nets were cleaned and stored in the canoes. The many containers of dried, smoked salmon were loaded and it seemed there would be little room left for the people. In small groups they paddled out into the open water and rounded the northern point of Samamao. Tcawogs, the people called it. From there one could see the stockade growing across the straits at Temxwigqsan.

Cha it zit, now taking Wanana in his canoe, fought the currents of a riptide as he propelled the heavily laden canoe. Dried salmon, wrapped in bundles of cattail mats, took all the available space. All the other canoes, also floating low in the choppy water, were as fully loaded. Cha it zit felt satisfaction in knowing that this, his first food-gathering expedition, had been successful. Then his thoughts traveled to other concerns, running at random across the threads of his memories and desires.

"This is where my vision came," he mused. In his mind he could hear the voice that had said, "Your wealth will be in your house."

Across the water, the village of Temxwigqsan grew larger. He seemed to see a longhouse, his house standing there, a bit apart from the others. "That is where I want to be," he told himself. "That is where my *power house* will stand. From it, the wealth and prestige promised me by the spirit Tiolbax will flow into my hands. In it I will give many potlatches. Everyone will know the name of Cha it zit."

He glanced at his mother, nestled amid the boxes, baskets, and bundles, her back straight, her hair, black as a raven's wing, held securely in two braids. She had ceased to paddle, and was resting her blade across her thighs. Since his encounter with the medicine woman, Cha it zit had thought about Wanana, about her baby coming, her chance of a safe birth. He wondered how it felt to be a woman soon to have a child. Then he stopped himself. "Do men concern themselves with such things?"

"That is a woman's role," he decided; it showed weakness to become involved. Or did it? Cha it zit wrestled with the subject. If he were to become a father, as Sax ump ki was, how would he feel? How would he act? He did not know. There were taboos, foods a man could not eat, things a man could not do, so that the spirits would be pleased and cooperative with the woman. Everyone knew that. Maybe that was enough.

Wanana, too, was lost in her thoughts. Largely, they revolved around Cha it zit. It was not considered appropriate, when a son is almost a man, to cling to him. She must be strong and push him into a man's world. But there, the two of them alone, she felt so very close.

"My son," she said after they had passed beyond Temxwigqsan and were paralleling the mainland, running close to the shore. "It is pleasant being here with you."

Surprised, his spirits rose when he realized that here, at least, they could talk. "I am glad," he replied, searching for the right words. "I have wanted to . . . to exchange . . . to talk to you. It's been – "

"I know," she helped him. "It's been hard for me too. But that is the way it has to be. You are in a man's world, now. But I want you to know that I am proud of you."

The village began to fill up with the returning people drifting back and they began unloading their canoes and settling back into their permanent homes.

Tseutsi's wrinkled, brown face stretched into a toothless smile when Cha it zit and Wanana entered. Old Sta nek rose from his pile of robes and hobbled toward them, his arms outstretched, "It is good. You are home," he cackled. Before long the others began to arrive. Bundles were stored and the big old house was filled with activity.

That evening, Cha it zit recited the story of the defeat of the Yakultas to the great delight of Sta nek. He wanted to know all the details. And he slapped his bony thighs, laughing with delight over each incident.

Several days later, Sax ump ki and Cha wentz arrived with the news that the fortification was almost completed.

"It is time for us to go to the island Swelax," Sax ump ki announced. "We will make our camp near our traditional village of Allulung. There we will gather more food: clams, sea mammals, and deer. We will need much food, for this winter there will be great demands on our supply. We have obligations coming upon us."

The people knew what he meant. Latsi would be reaching womanhood soon; she would need a great name and much food would be needed for the feast to be given.

Cha it zit had begun to realize a truth. In working shoulder to shoulder beside the fishermen of the village, in watching the endless labors of the women, he came to see that food was power. A man's success, he saw, was measured more by the store of food he amassed than by all of his other possessions. With a surplus of food he could give great feasts and thereby gain recognition. He could give food to those in need, and gain gratitude and respect. "The greatest gift a man can have," he mused, "is the power to get food, for it leads to all that it is good to possess." But quickly he checked that thought. "No," his mind said, "your gift, to succeed by gambling, is the greatest."

"Still," he thought, "if he could have both gifts, how great he could become."

Once again the people prepared to take down their mats and load their supplies into canoes and travel on their quest for food.

9 Spirit Canoes

Allulung, the ancient village site, lay on the far side of the island of origin of the people. It was only a heap of moldering cedar planks and a few upright giant logs protruding from the evergreen forest above a deep, narrow inlet.

Cha it zit sat back, eased his paddle onto his knees, and let his canoe drift slowly toward the sandy beach at the head of the bay. Others had arrived ahead of him – their canoes lay along the shallow beach. Already the women were setting up temporary shelters, mats on frameworks. Soon they would take their digging sticks and begin the arduous task of digging a mountain of clams: steamers, cockles, butter clams, and giant horse clams.

The men would resume fishing, trolling for other salmon tribes, as well as taking halibut, cod, skates, or whatever bounty nature offered them, but the women knew that important as the fish were, it was the clams that often made the difference between plenty and famine.

Day after day the women labored, backs bent, at the clam digging. Then they steamed them, cleaned them, strung them on skewers or twine, and smoke-dried them.

The work was pure drudgery and it seemed to be endless. If there was any extra time, they picked berries, dug fern roots, or gathered basketry materials. They did it all without complaint, for that was the way it was.

One day, one unforgettable late spring day, a strange thing happened. The men had returned from their fishing and were unloading their catch when someone shouted, "Look! Look at what is coming into the cove."

"Is it real?"

"It's a spirit canoe."

"A vision."

The people were frightened, excited. Cha it zit raced toward the beach, drawn by this phantom ship. He had heard stories of those spirit canoes. Tselique had spoken of them. Was he actually seeing one? It came closer. The white spirit blankets full of wind seemed to be drawing it to them. Then he saw beings running on the craft, the blankets tumble down, and the canoe slowed, but drifted closer.

Before long the large canoe gave birth to a smaller one. The beings got into it and began to propel it in a strange way toward the beach. Some of the women hid in the woods but Sax ump ki, Cha wentz, What la cum, Cha it zit and some of the other villagers stood quietly waiting.

The creatures, man-like beings with white skin, bushy beards, and strange garments, stepped ashore and stood facing them. Several brought a round box and a large bag, which they sat on the sand. They made sounds in a strange language. No one could understand what they meant.

Then one of them held up the pelt of a sea otter. The people wondered about that as they waited quietly and respectfully. From the bag he took out a handful of shining beads, more beautiful than the prized abalone shells. The people gaped at the sight.

Sax ump ki said, "I think they want to give us those beads for sea otter skins."

"Yes," What la cum agreed. "But we don't have any. We could offer them fish."

Sax ump ki took a large salmon from his boat and held it up. The beings shouted something – no one knew what, so the people waited and wondered.

Then the spirit, if that's what it was, opened the round box. He dipped some liquid from it and offered it to Sax ump ki.

Because it was proper to do so, Sax ump ki took the dipper and drank from it. All the people watched. They were horrified when their leader grabbed his throat and gagged. His eyes bugged out and he gasped for breath. Cha wentz leaped at the men, knife in hand, but Sax ump ki stopped him.

"No," he shouted, "no, do not fight them. This is strong medicine. What they offer us is powerful. We must have some."

Then he took all the fish from his boat and laid them beside the box. The strange visitors all spoke, but no one knew what they said. Sax ump ki looked around, saw an almost new blanket that Klatok had made for him, and lay that beside the fish.

The being held up the sea otter fur again. This time Sax ump ki gestured, "We have none."

The men took the fish and the blanket and put them in their small canoe. They kept the beads, but they left the box. Soon they were gone, into the large canoe, and the people were left standing on the beach, astounded.

For a while no one spoke. Then they all spoke at once. They congregated in groups discussing what had happened. Some were convinced that they had seen a vision, others said that the aliens were supernatural, a few declared that they were really people, like themselves, but disguised. No one could explain the "medicine" that the strangers had left. They expected it had magic power, the power to induce a trance or altered mental state since, after drinking of it, Sax ump ki had begun to act very strangely.

Cha it zit was frankly astonished at the way his father was behaving. While Sax ump ki was usually very serious, as befitted his high status, and carried his rather large muscular body with dignity, he was now laughing at nothing in particular and even tickling several of the younger women, including Cha wentz's wife.

The afternoon faded away, the sky and sea darkened, but no one thought of starting fires for the evening meal. A great, even epic, event had occurred; who could think of food? The people murmured among themselves not knowing what to make of the visitation. Should they be afraid? Should they leave this place, their ancestral home, where they had a right to be, or should they stay? They stared at the round box. Was it a threat or was it a *spirit gift*? No one knew which. Nothing in their lives, or in the remembered lives of those who had gone before them could have prepared them for the events of this day.

Sax ump ki wandered over to the round box and removed the cover. He dipped a clamshell full of the strong liquid it contained and smelled it. He dipped his finger in it and touched it to his lips, then he took the least sip and let it trickle down his throat. He smiled. "It is power medicine. Only those with great strength should drink of it, as I did. A warrior might manage to contain its power." He handed the filled shell to Cha wentz who, to show his own power, drank it down. His face turned red as he choked on it. The people laughed.

That was when Klatok spoke. "We will use it as we use the oil we prize. We will dip our berries in it. Then all may benefit from this new spirit medicine."

Sax ump ki nodded in agreement.

The women and girls had picked fresh salmonberries that day. Those were dumped into trays and then the golden brew was poured over them. Then the women, using their pestles, mashed the concoction together.

Cha it zit watched the proceedings and debated whether or not he wanted to partake of the medicine brew. He saw others, all proven warriors and successful hunters, ladle portions from a tray and warily sample them.

"It is powerful," they agreed, grimacing and clutching their throats. "Still, it is pleasant," they conceded later, smiling. Others took courage from their example and tasted the mixture, then came back for more.

Before long Cha it zit, who possessed no warlike spirit to protect him, felt obliged to at least taste the strange liquid. The bit he took on his tongue burned and when he hastily swallowed it, it seemed to sear his throat, which tightened, causing him to gag. "Yes," he admitted as the others had done, "it is strong."

New sensations, born of the liquid, arose in his body. He felt warm, peacefully warm. And the warmth radiated outward from his belly. That he liked. So he took more of the brew. It was easier to swallow the second time. The warmth reached his brain, "Ah," he thought, "this is good." A few more draughts and a feeling of euphoria settled upon him.

"They must have been spirits," he reflected upon the strangers. "Only supernatural beings could bring such a gift as this." Sometime later, and after more consumption of the liquid, Cha it zit began to feel dizzy. "This is strange," he admitted. He tried to rise but his body refused to respond, so he settled back.

The scene before him came into focus, became hazy and confused, and then cleared momentarily. He saw that the others were lying about, as he was, many with silly smiles on their faces. Some were laughing at nothing in particular, it seemed. Some were singing tribal songs together, or, he thought, attempting to sing songs together, for the result was a cacophonic disaster. Far into the night, the people ate the "medicine" soaked berries and reveled in this new experience.

The next day found them drooping, moaning about bad spirits roving through their heads. That was when the medicine woman spoke her mind. She had remained quietly in the background all the time that the people were discussing the contents of the round box. She had refused the soaked berries and watched the results of the orgy from a distance. Now she would have her say.

She flounced into the assemblage of suffering people and shouted at them, enduring Sax ump ki's ugly scowl and her own husband's look of displeasure. "You people are fools," she said. "Look at you. That is not 'medicine' as you think; it is poison. If you drink too much of it, you will die. Those who brought it were not spirits, they were men like all men. They are evil. They bring pretty things, yes even useful things, but they demand much for them."

What la cum moved toward her to stop her but she whirled on him. "I speak the truth. Are you afraid to hear the truth?" she shrieked. "Where I came from, the S'klallum people, there have been many men, and ships, too, like the ones who came here. They brought the drink to our people, whiskey, they call it. Now many of our men want only to drink it. They hunt sea otters, which is what the strangers want, to trade them for whiskey. The people go hungry. Don't let that happen to you. I warn you, don't let it happen to you."

"Woman, you forget your place," What la cum said, pushing her from the circle of people. She ran off, and disappeared into the fringes of the forest. What la cum followed after her.

No work was done that day. Sax ump ki replaced the cover on the whiskey container and stowed it in his canoe. Cha it zit, his head aching intolerably, wandered off toward the old village site.

The village of Allulung had been built on a rise above the beach, as was customary. Between it and the shore, a long mound snaked along the contour of the ground. Cha it zit recognized it for what it was, a midden, or garbage heap, where broken implements, clamshells, and discards of all sorts were tossed. He took a stick and poked into it in a desultory manner. His head throbbed, his whole being ached, so nothing interested him much.

In his stabbing, he unearthed a small bone object. After rubbing off the accumulated ashes and dirt, he found that it was an amulet, a carved salmon.

"It didn't help its owner," he thought, "he lost it in a garbage heap." Then he examined it closely. It was finely worked, unlike any that he had seen. "Yakultas," he reasoned. Only the hated Northern People carved like that. "It might have power." He started to put it back, but something stopped him. "It could be a gift." So he put it with his other charms in the leather bag that he always wore suspended by a thong around his neck.

There was nothing left of the village except the old, rotting framework of the houses, so he wandered up the slope behind it to a high open space. There he looked out across the inland sea, dotted with islands. The water changed from green to blue to slate gray, as dozens of currents cut through the passages between the islands, meeting in tide rips that boiled and surged. In the distance, in a small bay, he saw the white wings of the two spirit canoes. He could barely make out beings on the shore. It looked as if they were taking water from a creek that flowed into the inlet. "Do spirits drink water?" he wondered. The question hit him with a powerful impact.

Were they spirits? If not, they were men, just as the medicine woman had said. If they were men, men different from any he had known, where had they come from? How was it possible? All the learning of his clan said that the people came from here. "We are the people," he said out loud. Worse yet, there were two canoes, how many more were there? The medicine woman had said many. Should he speak of his fears to others? Would they think he was afraid? Weak?

He ran back to where there was comfort in the presence of his people. He couldn't speak of what was in his mind, but he could be with them, be where things were as they always had been. But were they? His aching head, the strange feelings of the night before, the "medicine" itself had already changed their lives, he worried.

"Xedewa," he thought. "Where is Xedewa? She knows of those beings. I will speak to her."

The villagers were as he had left them, dragging about, nursing aching heads, red-eyed and miserable. He spied Wanana, Latsi and Tsexad on the beach, baskets at their sides digging for clams. Putting custom aside, he walked toward them. They seemed to be as always, cheerful and busy.

"Cha it zit," Latsi called out happily, "do you feel the 'medicine'?"

"It is strong," he told her. "It is beating itself against my skull. Don't you feel it?"

"Xedewa wouldn't let us touch it," she said. "Isn't it awful what it did to everyone? She is right, you know, it is poison."

"Where is Xedewa?" he asked.

"We can't find her. I think that What la cum beat her. He was angry that she spoke out like that."

"Where are you going?" Latsi called out after Cha it zit, who was running away from the campsite.

Neither What la cum nor Xedewa could be found. No one had seen them since he had flung her from the circle of people. Cha it zit followed the route they had taken, tracking them, checking for bent branches and trampled ground.

He wasn't sure why he was meddling in what was strictly a family affair, a man chastising his wife for assuming to reprimand his own actions and those of his family. "What la cum has a right to punish her," he conceded. "She is a headstrong woman; she needs to be corrected. It is not my affair." But arguments had no effect on him, and he kept his course. "Maybe it is the 'medicine,'" he told himself. "My head isn't thinking."

At last he found himself climbing. Up, up, he went. Then he came to a small lake, a perfect pool, like a piece of sky fallen to earth. There, lying half in the water, he found her. She was still, her black hair floating beside her. There were blue spots on her arms and a gash, oozing blood on a leg. He raised her head; it was warm and her eyes opened.

"Xedewa," Cha it zit called her name. "Are you hurt badly?"

"How did you find me?" she murmured. "I thought I would die – die."

"Let me help you – get out of the water."

"My hurt isn't only of the body," she assured him. "It is also of the mind."

"Where is your husband? Did he do this to you?"

"What la cum? Oh he tried, but I scratched him and got away. I ran and ran. No man is going to abuse me again, ever."

"But the lake – why are you in it?"

"I don't know; I suppose I slipped and fell. I don't remember. You haven't seen my husband?"

"He hasn't come back."

"Maybe he is ashamed to show the scratches, to show that his wife could beat him?" She laughed sarcastically.

"We should go back," Cha it zit said "Can you walk?"

"Let me stay for a while," Xedewa asked. "It will be hard to go back now."

He pulled her to her feet; she reeled slightly, and he could see a nasty bruise on her forehead. They found a mossy log and he eased her onto it. His own hurting head rebelled at the effort.

"I want to know about those beings," he told her. "You said there were many at your village. What did you mean?"

"The first ones came a long time ago, before I was born. My parents told me of how two of the large canoes with white wings came from the great open waters, through the sea channel to our village. They did not harm the people, they only came by. We traded them some furs and fish for beads, beautiful shiny beads, and good things like tools and blankets. Our people thought they were spirits then and respected them. We treated them with honor, and they treated us the same."

"Were they the same as the ones who came here?" Cha it zit asked.

"In some ways they were different. The headman, his name sounded like 'Captain Quimper' and he wore clothing of the most delicate nature. His hat was most unusual, not woven of roots as ours are, but of some sort of wool or fur. Large feathers, unlike any that our birds have, decorated it. He had facial hair, so much that he looked like a goat. It was dark, too, and curly. Those other men who came here had beards, too, but most were rough, they looked more like sea lion bristles. Their clothes were rough, too, not at all like Captain Quimper's were. Those men are like the ones that come to the S'klallums for sea otter furs. We call them 'Bostons' because that is the place they claim to come from. They are like the men who came here."

"Bostons." At last Cha it zit could put a name to the so-called spirit people. "That Captain Quimper, he came a long time ago. Did others like him come, too?"

"So I believe. Every so often someone would see a ship go by. That's what the Bostons called them, ships. Then, when I was almost grown, two ships came. They stayed near our village for a time."

"Did you become friendly, them and your people?"

"Oh yes, as I recall, the name of their headman was Captain Vancouver. They were not very interested in furs, but they did trade good things like knives, axes, cooking pots, blankets, and beads for almost anything we brought them."

"Why did they come?" Cha it zit asked. "Do you know? What did they do? Where did they come from?"

She looked perplexed. "I do not know those things. I only know that they came. Each day they got into smaller ships, they called those 'boats,' and they went away for a while, but I don't know what they did."

Xedewa put her hand to her bruised forehead, felt the swelling, and stared at the drops of blood that came off on her fingers.

Cha it zit jumped up. "I have been unfeeling," he said quickly. "You are hurt and I ask questions. Let me see if I can find some of those spotted leaves to put on that wound."

He reeled a bit, his head still throbbing, but he started up toward the dark forest that covered the upper slopes of the mountain. She didn't restrain him, as he had half-hoped she would.

It took courage to venture alone into the dense growth of cedar, fir and hemlock. Shafts of sunlight penetrated the upper layer of sun-seeking branches, but a dank gloom lay over the beds of sword ferns that carpeted the forest floor. Here and there, where there was a bit of light, vine maples sent out inquiring shoots which snaked along the ground or spiralled into the air, seeking the life-giving sunshine. In this world, apart from the sea, forest spirits dwelled, and evil beings that could suck the life from a person, or could drive them mad, or carry them away to be devoured in some hidden lair. There were animals, too, who could easily overcome an unarmed man-boy, but Cha it zit chose not to think of them. He concentrated on his task, finding the healing leaves.

In a damp place near where a spring oozed from the ground, he found several small plants bearing the same mottled, juicy leaves that Xedewa had used to heal the scratches she had inflicted on him.

When he returned, she was lying on the ground staring up at the sky above the small lake. He tore the leaves apart and placed them on her temple.

"I didn't think you could find any," she commented. "You must have a special gift for healing. The plant gave itself to you, or you would not have seen it."

Cha it zit was about to tell her of Tselique and the gift the shaman had given him, of the boards and the other magic for healing which now belonged to him. But he checked himself in time. To mention Tselique's name and even hint at the things Tselique had told him would be to open himself and his clan to a terrible disaster. Instead, he said, "To heal is a gift. It interests me. It is good to help our people, but I do not choose to be a shaman."

"One does not choose," she corrected him, "one is chosen. I believe you are to be one of those. If you wish, I will show you what plants are medicine and how to use them."

"I have a spirit, a different one. I have the one I sought. That is enough."

"As you will," she replied in a formal tone. "Now it is time that you return to the camp. The people will miss you. I do not wish them to find you here."

"But I came looking for you, to bring you back. You cannot stay here."

"I will not go back."

"You won't be safe here. The evil spirits, the animals – "

"It is better for me to stay here. Now go."

"Then I will come back. I'll bring you food."

When Cha it zit entered the summer camp area, it was obvious that the men had been dipping into Sax ump ki's round box. They had carried it to the beach and were passing ladles from one to another, sipping of the strong whiskey. The women sat sullenly on the outer fringes, watching their men, and showing no interest in preparing the evening meal. Wanana was not with them, nor was Latsi or the slave girl, Tsexad.

Cha it zit was worried. He thought, "That is bad medicine; already it is dividing the people. It is taking away their pride, their ambition. It is to be avoided. I wonder if that will be possible."

What la cum, who had skulked back to the camp while Cha it zit was gone, was getting boisterous. He was drinking deeply from the ladle each time it was passed. "Where is that accursed woman?" he shouted, standing and reeling awkwardly.

"Sit down," Sax ump ki laughed at him. He was drunk again, too. "She is only a woman, let her be. She will show up soon enough."

What la cum sank back down again.

Cha it zit was hungry. He looked for Wanana. Surely she would find a morsel for him. Walking toward the beach, he saw the lazily drifting smoke of a fire. There he found them, Wanana, Latsi, Tsexad, steaming the clams they had dug.

"Isn't it terrible," Latsi asked, "the way the people are acting? What makes them so ugly?"

"The bad medicine," Cha it zit answered.

"Have you seen Xedewa?" the slave girl asked, her eyes downcast.

Cha it zit hesitated, but seeing the worry in her eyes, he told her where her mother was, and what had happened.

“I will go to her. She may need me,” Tsexad said rising to go. “Please take me.”

“No, she is afraid What la cum will find her. He is in a bad mood and he might follow you. I will take her food since he will not suspect me.”

The pit in which the clams had cooked was opened and the aroma invited them to eat. When Cha it zit had had his fill, he gathered the leftover clams into a large maple leaf and stole softly out of the campsite.

It was dusk when Cha it zit reached Xedewa’s hiding place. His head no longer hurt and the clams had given him energy. Xedewa, however, had not fared so well. She had taken her cedar bark skirt off and was lying under it on the ground. He handed her the clams, which she ate avidly. Neither of them spoke till she had finished.

Then she studied him. Her intense look bothered Cha it zit. He moved constantly, walking from place to place, but her eyes followed him. He became uneasy. If this strange woman really was a shaman, as she had said, she could put a spell on him simply by staring at him.

“I must go,” he said. He looked up at the open sky above the lake. Already stars were showing.

“No,” she said. “It is too dark. You must stay.”

“They will miss me.”

She laughed, “Miss you? By now the men will have finished the whiskey. They will be too numb to miss anyone. You must stay.”

She was right, he knew.

Again she regarded him and then she asked an unthinkable question. “Nephew of my husband, have you ever been with a woman?”

Cha it zit was embarrassed. Whether it was because the answer would be “No,” or whether it was because he was afraid of what she would say next, he did not know.

“You do not answer. Is it because you have not?”

"Yes," he said.

"Among my people," she continued, "there is the custom that a married woman can be given to a near relative of her husband if that person asks for her. Is that your custom also?"

"It is," Cha it zit answered. "All the members of a family share. That includes wives."

"Then," she said, "it is time, and past, that you know a woman. There are things you must experience. Things that only a woman who has the right knowledge can show you. As your uncle's wife, I offer you that experience."

"I have thought of it – what it would be like – but I'm not sure – " he began, embarrassed, and yet somehow eager.

Xedewa tossed off the skirt and held out her hand to him. "Come," she said softly.

Her body was fair. She was older than he was but time had dealt gently with her. Her breasts were firm, her flesh rounded. Cha it zit took her hand and she pulled him down beside her.

She caressed him and a warmth grew in him until it burst into a conflagration that consumed him. He forgot everything except himself and the woman's body. She paced him, matched him movement for movement. Never before had he experienced such sensations. At last, in a burst of feeling, it was over. He lay on the ground, spent but curiously fresh. He felt renewed, alive. Xedewa rose and walked to the lake to bathe. When she returned, she wrapped her skirt around herself and regarded him calmly.

"What are you thinking?" she asked him.

"That I will never be the same again," he said.

"No," she agreed, "you never will. Now we must sleep."

In the morning at the first rays of dawn, Cha it zit left, telling Xedewa that he would return in the evening with food.

"Bring my daughter," she had called after him.

The people were just stirring. As he had expected, the men still lay on the ground, the empty whiskey box lying on its side, dippers here and there. The women, hungry by now, were stirring the ashes of their fires, feeding tinder of dead cedar twigs, coaxing flames.

Cha it zit loosed his canoe from its moorings and paddled down the long inlet. He had no destination, no purpose, no plan. He only knew that he must be alone to consider the events of the past several days. It was as if all the challenges, all the defeats and successes of his life had been leading to this time. He knew, instinctively, that in those past hours he had become a man. Not gradually, but suddenly, one act had told him he was an adult. In his mind he experienced again the delicious excitement of the night before.

He knew, too, that the white spirit beings of the legends and of the elders' stories were men. Different, but men like himself. What he didn't know was how those aliens would affect his life and that of his people. If the whiskey was any indication, he thought, they would bring trouble.

The morning mist was rising from the water. Faintly, the sun shone through, bringing warmth. Cha it zit lifted his paddle, laid it beside him, and let the canoe drift as it would. There was no sound except the gentle gurgle of the water as it slid beneath the craft. Off to the right, near a slice of mud beach, a pair of herons stalked their prey. He watched as they each stood on one incredibly long, thin leg, motionless, waiting for a small fish or a shore crab, or any edible thing to come their way. Then, with a strike so fast that the eye could not follow it, the catch was made.

His canoe slid onto the tide flat and stopped. Half open cockle clams lay partly covered by the sand. He gathered some and watched as they snapped their shells shut. With his shell knife, he deftly cut the muscle and the helpless clam fell open.

"If the herons can fish, I can too," he decided. So he continued out toward the strait that separated the Island of Origin from another low-lying island. There he baited his hook with a clam and let the line deep into the water. The fishing was only incidental, an excuse to drift, just drift.

To his surprise, there was an immediate strike. Almost effortlessly he pulled up a good-sized halibut. When he gaffed it, it seemed to help him lift itself from the water. It lay in the boat gasping until Cha it zit stunned it with his fish club. It happened again and again – the fish seemed to want to be caught. He couldn't understand it. "I never was a good fisherman," he said. "Now suddenly they are coming to me." He was sure he had somehow gotten a spirit helper, the kind the fishermen get.

Then he remembered the amulet he had found at the old village site of Allulung. Could it be that he had received so great a gift? That charm, waiting there for him, one that had been the possession of a fisherman long ago, was all the evidence that he needed to convince himself that he had indeed received the spirit help that he needed to be a catcher of fish. He marveled, inwardly, at the things that were happening to him.

Too soon, his canoe was loaded, overloaded, with the morning's catch. He turned and headed toward the summer camp. By that time, the men were up and preparing to fish. They showed the after-affects of their bout with the whiskey, moving slowly, bleary-eyed. When they spied Cha it zit returning with his catch, they called out, "See, the cub has become a he-bear. He has learned to fish."

"Yes," Cha it zit agreed affably, "I have become a he-bear."

10

Younger Brother

Cha it zit landed his canoe and walked to the place where Wanana and Latsi camped. They were skewering horse clams on twisted cedar bark twine, preparing to dry them. Bent over their task, they didn't notice Cha it zit. He regarded them. He saw that Wanana, now enlarged with the child in her belly, looked tired. There were circles under her eyes and her mouth, usually soft and tender, was set in a grim line. When she moved, it was with difficulty. Her skirt, fashioned in the way all women made them – shredded cedar bark, woven at the top, hanging freely over the thighs – was frayed from the constant work she performed.

"Her time is near," he thought. "Then she can rest."

For the first time he felt pity for his mother, for all the women. The feeling was strange, particularly for a man, and he dismissed it.

"Women are born to work and bear children, it is their duty," he assured himself.

“I have brought many fish,” he announced, an unconscious pride revealed itself in his tone. He spoke in a low voice, as befits a man.

Wanana jumped in surprise and looked at her son. He returned her look, something he had not done since his spirit quest.

“I will help bring them here for you to prepare for drying.”

Wanana, made a helpless gesture. “Go with him Latsi,” she ordered. “I will work with the clams.”

The two hauled the catch, dragging the heavy halibut up the beach to the high ground.

“I am hungry,” Cha it zit said.

Wanana rummaged through a basket and found a piece of fish left over from a former meal. She handed it to him.

“Where is Tsexad?” he asked, casually. Neither of them had seen her all morning.

“She is probably looking for her mother. We haven’t seen Xedewa since What la cum chased her away two nights ago,” Latsi answered.

“If you see her, tell her I want to talk to her,” Cha it zit requested, mumbling between bites of the fish.

“Why – ” Latsi began. Wanana interrupted her.

“That is not your concern, daughter. We will do as your brother asks.”

Cha it zit went back to his canoe. The other men had launched their craft and were far out into the inlet. He followed. Again the fish swarmed to him to be caught. He marveled. He removed the amulet from his pouch and fingered it. It felt warm. “It must be the magic in the charm,” he decided. “I have a fishing spirit and I am a man.” He rejoiced.

Toward evening, he and the other men returned with their catch. Tsexad stood on the shore waiting for him. She helped him carry the fish he had caught to the camp. Then he asked, "Do you have any food?" She had procured some berry cakes and steamed clams. "Come with me," he ordered her.

When they arrived at the lake, Xedewa was waiting for them. Tsexad gave a little shriek of delight and ran to her mother.

"I looked and looked for you," she laughed.

"Come," Xedewa replied, "we have much to discuss. You and me and this man." She smiled at Cha it zit knowingly. He grinned and then wondered if he had reddened a bit.

They made themselves comfortable on the mossy ground overlooking the lake. Flies hovered over the water, and now and then a trout would leap for its dinner. After the intense work of the day, it was peaceful to just sit and feel part of the land, the sky and the water.

"For many days I have thought," Xedewa began. "My position in this family is a weak one, the small third wife of a second son, but I am content. What la cum will not inherit position or wealth when Sta nek dies. That which has not already gone to Sax ump ki as oldest son will be his when the old man goes. The same will be true for Cha it zit," she said, turning directly toward him. "Cha wentz, as the first born, will be headman, and none of the family wealth will be yours so long as Cha wentz lives."

"Why do you tell me this?" He was indignant that this foreign woman would discuss his family heritage, would put him in such a position.

"I do not mean to offend you, but hear what I have to say." He was silent. "My daughter is of good lineage. There are no greater S'klallum people than mine. Her father was a warrior of noble rank among the Yakulta people. She has strong blood. Her rank is slave only because my lover was a prisoner, captured in battle. But what will become of her? What future does she have?"

Cha it zit could not answer, although he knew.

Xedewa continued. "She will be available to whichever of the men wants her. I cannot protect her, for I have no status with your people."

"Why do you talk of this to me?" he asked again.

"Because," she said, speaking slowly and deliberately, "I ask that you take her."

"Me?" he was astounded.

"You. Not as a wife. Even a second son must marry his first wife for her social standing, to bring prestige to the family. Take her as your slave. More than a slave, as your helper, your companion, your possession. You will be a great man someday, that I know. Keep her safe and be kind to her."

"How can I? I have no house, no possessions."

"You will have. Is she not beautiful?" The mother held the girl's face between her hands and turned it toward Cha it zit. Tsexad's eyes were wide, frightened as a fawn's, and as soft. "If it was not for her round slave's head, she would be more noble appearing than any of the women in your village."

Cha it zit glanced at her. She was good to look at. Her skin was paler than the others, her brows were wide and arched, and her mouth small but full. He had not really looked at a girl before, not like that.

"I have had an offer of marriage from the headman at Temxwigqsan," he ventured.

"Then this will be an advantage," Xedewa said. She would not be put off. "Slaves are of value. It will increase your standing."

Cha it zit nodded. He could think of nothing more to say. Tsexad, eyes down, was silent, too.

"It is done then," Xedewa said with satisfaction. "I have one request of you, Cha it zit. If you desire my daughter, remember what I taught you last night. Treat her as I treated you, tenderly. She has known no man."

It was done.

The next day Xedewa slipped back into the camp and took her place at the drying racks, processing salmon, stringing clams. What la cum accepted her presence in a noncommittal way, but there was a look of relief on his face. If the villagers noticed that Tsexad remained close to Cha it zit, doing his bidding, they did not acknowledge it.

The time of the low tides had come. With the first warm, long days of summer, the sea ran out great distances to expose bottom sands that were covered most of the year. That was when the women gathered the large horse clams that lived deep in the sand at the farthest reach of the tides. With their fire-tempered digging sticks, they dug deep holes, working quickly before the huge bivalve could draw his neck down a foot and use it to dig another foot deeper into the sand.

The clams were tough, but the meat was sweet and tasted of the sea. When they had been smoke-dried on spits, they were like wrinkled bark; but pounded and made into chowder in the winter, they were delicious and nourishing.

Day in and day out the women labored on the clam beds. After they had packed away enough to feed their families for the winter months, they processed more for feast-giving and trading. They knew that more important than prestige or wealth, or spirit power, was food. Everything depended on plenty of food and that depended largely on the work of the women. So, without complaining, they labored on.

Meantime the men turned their energy to hunting deer. Having been trapped on the island generations ago, inbreeding or lack of browse, or perhaps both, had resulted in a herd of pygmies. The tiny deer were more easily caught by running them into a net strung across a narrow spit than by spearing or using bows and arrows. For as long as the

tribal memory could recall, this netting of deer had been done in one place.

Cha wentz and Cha it zit had strung the heavy cedar branch net, concealing it with foliage. They crouched well back on either side while the beaters, Sax ump ki, What la cum, and others from the village beat with sticks and whooped, driving the frightened animals toward the net.

When the first of the deer, an antlered buck, ran into the net, breaking his neck, Cha wentz leaped to pull it aside lest the rest see his fallen body and bolt back the way they came. As each deer arrived, it was taken and quickly dispatched with a well-aimed blow to its head. At day's end there was a heap of carcasses, already gutted and skinned, to be carried to the tired women.

Turning from their endless digging, they cut the venison into strips and dried it in the sun. Smoky fires discouraged the ever-present flies, now grown to great numbers, drawn by the discarded bones and offal of the fish and the shells of the clams.

The skins were wrapped in bundles and tied tightly to be processed into leather at a later time.

Day after day, without ceasing, the people worked. They moved from one task to another, depending on the need. If the fish failed to show, the men went for seals, or sea otters, or even porpoises. The days grew hot and then hotter.

The people ground a red clay, rich in iron, with fish oil and rubbed it on their skins to prevent sunburn and chapping by the salt water. They wore woven hats with wide brims to protect their heads and eyes from the heat and sun. The men often wore no clothes at all, while the women shortened their skirts.

In the heat of the hottest day, Wanana fell to her knees. Her boning knife tumbled to the sand and she clutched her swollen abdomen. Xedewa, who was working nearby, ran to her, followed by Latsi. "Her time has come," she told the girl. "Take mats to the meadow." Xedewa pointed toward a clearing among a stand of madrona trees. "You will find a framework there which I built. Tie the mats to it." Latsi hurried to obey and Xedewa ran to fetch her medicine bag which she had recently filled with medicinal plants.

Wanana had recovered and was attempting to stand when Xedewa returned. "Lean on me," she said, helping Wanana toward the meadow. By the time they arrived, Latsi had finished tying the mats to the framework, creating a small lean-to, and she was placing moss and ferns on the ground inside.

As she sank down upon the soft moss, Wanana stifled a groan. Xedewa took a bundle of dried Indian plum leaves from her bag and handed them to Latsi. "Boil these in water and bring the brew as quickly as possible," she ordered. Latsi left, running, as Xedewa called, "There are plenty of coals left at the fire pit."

Then Xedewa removed a stout cedar bark rope from a cache where she had left it a few days earlier. She tossed one end of it over the bough of a madrona tree that hung over the lean-to. Then, slipping the dangling rope through a loop she had tied in the other end, she pulled the rope tight and let it dangle into the birthing hut.

That done, she entered the hut and sat beside Wanana. Gently, she stroked Wanana's hair. "It won't be long now," she said. "I have attended many births among my people, you must trust me. Do what I tell you to do, even if it may be different from your customs."

Wanana, perspiration dripping from her body, contorted with pain, smiled at her wanly. "I have given birth before," she gasped, "that you know. But this is different. I try not to feel the pain, but it is great."

Latsi returned with a wooden bucket filled with steaming tea.

"Drink this," Xedewa ordered, giving Wanana a steaming dipper of the aromatic brew. "It will ease your pain."

There was nothing to do but wait. The birth pangs came with increasing frequency. They rolled like a storm through Wanana's tired body, but not a sound, except her heavy breathing, came from her.

When it was time, the medicine woman tied a cord of twisted inner cedar bark around Wanana's waist above her swollen abdomen. "That will keep your baby from going the wrong way, toward your heart," she explained to Wanana.

"Now," Xedewa said, "grab this rope. Pull yourself up on your haunches, it is the easiest way." Wanana struggled to obey and she pulled herself up into a crouch. Xedewa placed a bundle of moss below her. "This is for your baby when it comes. Now, pull of the rope and push hard. Help your baby to come."

Wanana tried. Over and over she pulled on the rope and she pushed until she felt that she could push no more.

Xedewa bathed her tear-stained, hot face with cool water. Then she heard Latsi's voice, "They are coming!"

Sax ump ki, with Cha it zit following, came running. Both were panting.

"How does it go?" Sax ump ki called, not daring to look into the hut. Only women or shamen skilled in childbirth could enter.

"It is a hard birth," Xedewa replied in a worried voice. "I fear that you have not kept the taboos. It is not going well. Maybe it was the whiskey that offended the spirits."

Sax ump ki did not answer, only Wanana's labored breath could be heard. Then, suddenly, unpredictably, she screamed. It was a strange, unearthly sound. Cha it zit tensed. He wanted to do something, but there was nothing possible.

"The baby is not in a good position," Xedewa called out. "I must use more than skill and herbs." She began to half chant, half sing. It was a strange, guttural sound, all the stranger coming from a woman.

Cha it zit was overwrought. It came to him like a thunderbolt that his mother might die. "I must do something," he told himself.

Just then the face of Tselique flashed into his mind. "What would Tselique do?" It was as if the vision looked at the leather bag that always hung from Cha it zit's neck. The raven charm that Tselique had given him long ago lay there. "You will know how to use this," the old shaman had said when he pressed it into the boy's hand.

Wanana cried out again. Cha it zit ceased to think. He acted. Taking the small carved bone amulet, he rushed into the hut.

Xedewa, still singing over Wanana, jumped back in surprise. Cha it zit picked up her song. It changed into another, one he knew he had heard sometime, somewhere. He held the amulet on Wanana's heaving abdomen. He could feel her violent contractions; he moved the charm as he sang. He seemed to feel movement from within her swollen belly.

Then she raised her head to look at Cha it zit, pushed mightily, and the baby came.

Xedewa eased the infant onto the soft moss and gently bathed it in warm water. The baby gasped and cried, a thin wail that seemed to fill the hut.

Wanana let the rope go and sank back on the ferns, now blood spotted, a tired smile on her face. "Thank you, my son," she murmured.

Cha it zit slipped out of the hut between the screen of mats and smiled at Sax ump ki. "You have a son," he said, "and I have a brother."

Sax ump ki gripped his shoulder. "You did well," he said. "I can see that you have great power."

Xedewa had work to do. She cut the umbilical cord the length of two finger joints. When the afterbirth came, she wrapped it in shredded cedar bark and handed it to Sax ump ki, who took it to bury in a secret place. Then she sprinkled cold water on the baby's back to make it strong. Reaching into her bag, she extracted a seal's bladder which contained dogfish oil. This she gently rubbed on the baby's skin, massaging the legs so they would not be fat, rubbing the abdomen to keep it small. She pushed down the sides of this tiny fingers to make them slim and tapering. Then she pinched his nose to give it a good shape, slender and straight. All of this was prescribed by custom.

When she was done, Xedewa took a piece of the softest fur, the pelt of a sea otter, and wrapped the baby in it. "What a beautiful child," she murmured as she showed him to Wanana who took him into her eager arms.

"Latsi, where is Latsi?" Wanana asked, her voice so tired, so soft, that Xedewa could hardly hear her. But Latsi, waiting outside the hut did.

She burst in, "Here I am, mother." Wanana handed her the fur-wrapped bundle.

"Give Sax ump ki his son," Wanana told her.

Latsi stepped outside, cradling the tiny bundle. Sax ump ki had just returned from his secret errand and she placed the baby in his arms. He looked at the red, wrinkled face, the mop of black hair, and then carried his new son to the campsite for all to see.

"I have a new son," he proclaimed.

"And Wanana?" the people asked.

"Because of my son, Cha it zit, she lives," he answered.

Cha it zit, who had accompanied his father, was embarrassed at what he had dared to do, but proud too. "She will be in good health," he said, "she only needs rest."

The people admired the new child and then regarded Cha it zit soberly. "Is it true," they asked, "that the child and Wanana live because of what you did?"

"What he did," Sax ump ki observed, answering for Cha it zit, "only a true shaman could do."

"I did what I was called upon to do. If I have shamanistic powers, powers which I resist with all of my being, then they must be used. I could not stand there and not try to help my mother in her time of need."

The people were quiet. They had witnessed a remarkable thing.

Overcome by emotion, by the implications of his extraordinary act, Cha it zit left the people. No one followed him, but they murmured after he had gone.

Xedewa placed a warm blanket over Wanana to keep her blood flowing. When Sax ump ki returned the child to Wanana, she received him happily.

"Is it well with you, wife?" Sax ump ki asked gently.

"It is well," she replied.

"I must attend to my duties as a new father," he told her, "but I will return to you and our son."

Wanana slept.

Xedewa kept watch over mother and child while Latsi waited nearby to do her bidding. "Bring me a fresh, skinned horse clam neck," Xedewa called to her. Latsi ran to do so. The medicine woman dipped her finger into the dogfish oil and placed it into the infant's mouth. He sucked it eagerly and wanted more. When Latsi returned with the horse clam neck, Xedewa gave it to the child to suck. He accepted it and closed his eyes, sucking contently. Soon both mother and child slept.

Later, when she awoke refreshed, Wanana steamed her breasts over hot rocks on which Xedewa sprinkled water. "The steam will make your milk rich and plentiful," Xedewa told her. "Latsi is digging clams for you and she is grinding mussel shells to put in your drinking water. Those things will make good milk."

Sax ump ki made a cradle for his son out of a slab of cedar. He worked it as one would a tray, making it as wide as his foot and almost twice as long. When it was finished, Wanana placed soft shredded inner cedar bark in it and placed her son there. She put tight rolls of cedar bark under his neck and knees and then placed a piece of mountain goat blanket over him. Then she laced leather thongs through holes that Sax ump ki had drilled in the cradle to hold the tiny body tightly in place.

Klatok, who knew about such things, was summoned to apply the pad of cedar bark which would mold the new baby's head into the proper shape for a high-born person. She came in briskly. She carried a small, flat stone and a thin slab of the bark to be used. She folded the stone into the bark to make a neat, flat bundle. She laid that on the child's forehead and bound it tightly, using buckskin thongs, lacing them through holes that Sax ump ki had drilled through the top end of the cradle. The baby cried, but that was to be expected. He was a strong healthy boy, he was merely asserting himself. Except for a daily bath, massaging with oil, and a change of the inner cedar bark bedding, the young child would stay on the cradle board until he was old enough to learn to walk, but the head-flattening rock would be removed after several months.

When the baby had been properly installed, everyone relaxed. All had been done according to custom and he was safe. Wanana's milk came and he accepted her eagerly.

"What shall we call this new one?" Wanana asked Sax ump ki.

He thought and then said, "I believe that this child would not be alive if Cha it zit had not acted as he did. Let us call him 'Soquwa' – Younger Brother."

"Yes," she replied, "I need to talk of that, of the birth. I don't remember much that happened, only the terrible pain, the effort. But I do recall Cha it zit coming in. It was an improper thing that he did. No man, except a shaman, may enter a birth house. Then I heard a song. It seems it was the one that another person, who I will not name, sang, and yet it was Cha it zit who sang it. Can you explain that to me?"

"No wife, I cannot. I will not allow myself to think what it could mean. We must not even talk of it. It could be dangerous."

"Will you ask our son? I fear for him. There is a power in him that he may not be strong enough to control."

"I know. I will do what I can. I can promise no more."

"Leave me now," Wanana said. "There is much for you to do. My little son and I will rest yet a while, but you need not stay."

The people had returned to their labor, the men endlessly trolling or sea hunting, the women forever digging clams or working at the drying racks. For weeks the racks had been full as the women rotated finished foods with fresh. As they worked, they spoke in careful, guarded words about the birth of Soquwa. They did not mention the name Tselique, the dead shaman whose song Cha it zit had unconsciously sung, but they remembered the old man's power. Had he come back to live again the in the body of Cha it zit? They avoided contact with Cha it zit, fearing that he might contaminate them.

"It is to be expected," Xedewa said, when he spoke to her about the feelings of the people. "You do have a shaman spirit, that is obvious. But you need not use it. Let the people see that. In time it will be as it once was. Well," she reconsidered, "nearly as it once was."

11

A Wife

Sax ump ki was satisfied. He looked down upon the encampment from a rocky shelf. The people, his people, were going about their tasks as usual. He saw the bales, baskets, and boxes of food piled high above the beach. It was more than enough to see the clan through the winter. There would be food to trade, food for feasts, and food to gamble with. The future was assured, for a year, anyway.

It had been a time of achievement, this food-gathering expedition, but there was more to be grateful for. The spirits had supported him in many ways. The protective wall at Temxwigqsan was finished, or nearly so. His new son was born and his wife, Wanana, was regaining her strength. His second son, Cha it zit, was showing himself to be a man, a man with exceptional powers. With such leadership as Cha it zit could offer, the future of his people was bright. All was well.

He was aware that the days were growing short. The time had come to leave this place; other tasks in other places needed doing. So he resolved to call a council and make plans.

Sax ump ki, What la cum, Cha wentz, and Cha it zit met the headmen of the other houses of Swetquem on the beach. After their pipes had been filled with dried knik knick leaves and lighted, they began to confer.

"The humpback fish will be starting up the Nooksack river soon," a grizzled old fisherman predicted. "There are traps to be made ready." The others agreed.

"We must take bears; we need the grease, and the meat will be welcome," a hunter said.

"I lack a bearskin robe," another announced.

"The mountains grow cold early. If we are to hunt mountain goats and the women are to pick blueberries and huckleberries, as usual, we must not delay," a sinewy man advised.

So the decision was made and the people began to pack their canoes. They now strapped boards across two canoes to make rafts on which to carry the vast amount of foods they had gathered.

Not as a group, but a few at a time, the people left, calling good-byes, and drifting out into the open water, the canoes and rafts clumsy under their loads. Gradually the camp became quiet. After many days living amid the noise and activity, Cha it zit found the near-silence pleasant.

Wanana, just out of her confinement, and her baby were preparing to leave. It was agreed that Sax ump ki would go ahead with the first canoes, along with Cha wentz and the other warriors, to scout for signs of Yakultas who might be waiting to prey on the women and children in their clumsy, heavy boats. Cha it zit was to escort Wanana, Latsi, Xedewa, and Tsexad. What la cum had his other wives to protect, and it was felt that Klatok, who was a match for any marauder, should escort the younger wives of Sax ump ki.

The family of Sax ump ki and What la cum was the last to leave. Looking back, Cha it zit recalled all that had happened on the Island of Origin. He watched the beach disappear, the trees grow smaller, and last, the ancient gray house posts fade from view.

What la cum took the lead, followed by the women. Cha it zit and Tsexad covered the rear. Their route lay along the edge of the island, then across open water toward the southern tip of Samamao, and then northerly into the bay that led to Swetquem. It was a dangerous trip. There were rocks and tidal currents at the jutting peninsula of Samamao that could easily swamp a canoe, tipping passengers and freight into icy, swirling water. There were times, too, when a wind would whip the bay into angry, spitting combers. All this the people knew and it was accepted as a part of existence. They knew, too, that if their spirit power was great enough they could control the elements and be safe.

Cha it zit accepted that he had spirit power. He had proved it, but whether he could control the sea he questioned. His canoe sat low in the water; it responded sluggishly to the thrust of his paddle. In rough water the load could shift, the canoe could roll, a wave could hit him broadside before he could swing the prow into it.

"Stop it," he told himself. "You must not be weak; don't think fearful thoughts. Think of other things."

He concentrated on Tsexad's strong back and arms. He admired the way she handled her paddle, each stroke true and sure. Although it had been many days since Xedewa had given her to him, they had spoken to each other only when it was necessary, and then the words had been few. There was so much work to do that the days had been spent in activity and the nights in a sleep of exhaustion. Their difference in status had kept them apart, too. It would be unseemly for a man to consort openly with a slave, and it would be presumptuous for a slave to expect it. But here, alone with her in a canoe, each dependant on the other, custom and tradition fell away. They dropped farther behind the others for no apparent reason.

"Tsexad," he said. She turned her head to see what he wanted, her long, black hair whipping across her face. "I just wanted to say that I am happy with you. You have worked hard."

"It is what a slave should do," she said bitterly.

He hadn't expected that reply. He was silent for a while, then said, "You are not truly a slave, you are only half-slave.

"What is the difference?" she asked. "I have no family to provide a potlatch. I have no wealth or prestige. What becomes of a slave?"

"I do not know. Perhaps something can be done," he tried to assure her.

They paddled in silence for a time. Then she spoke in a more congenial tone. "I am sorry, for what I said. If I must be a slave, I am happy to be yours. My mother was wise when she gave me to you."

"Speaking of your mother, she was telling me of the white men at your S'klallum village. I would like to hear more about them," he changed the subject. Tsexad half turned toward him, laying her paddle beside her.

"We have not seen many, but we have heard stories of great numbers of white men from our neighbors the Makahs, who live where the great open sea and the straits come together. Would you like me to tell you of them?"

"Yes, do. You rest and tell me what you know."

"First, let me explain about the Makahs," she began. "They are related to a people who call themselves Nootka, who live on the land across the great strait toward the north on the shore of the endless sea. They are a powerful people who hunt whales in the open water. They are fierce and possess much wealth. I know this because my mother's sister is married to a Makah who has a wife from Nootka."

"Yes, yes, but what about the white men?"

"I'm coming to that," she replied testily. "A long time ago, before I was born, the first white men came to Nootka. The Nootka chief's name was Maquinna. He was a strong chief. The people obeyed when he spoke."

"What did they want?"

"At first they didn't seem to want anything," she answered. "They had good things to trade, like knives, and axes, and beads. They took furs and fish in exchange. Later on, many boats came. They wanted the fur of the sea otter. They wanted all that the people could hunt. The Makahs say that the Nootkas grew rich from the trade for fur, and yet always the white men wanted more and more. They even built a village there and made their big canoes on the beach."

"How did they get along together? Were they friends?"

"At first, when the Nootkas thought the white men had special spirit power, they got along as friends and equals. But after a while there was trouble. The Nootkas tried to steal goods and the white men shot *thunder* at them. Many from Nootka were killed."

"Thunder? Were white men killed too?"

"Oh yes. They died the same as the Nootkas did. Chief Maquinna made slaves of the white men, too."

"They really are men, then," Cha it zit observed.

"Oh yes," she agreed. "And they are at other places, too."

"Where?"

"In the Tillamook country," she said. "That is south of the Makahs, down the endless sea, past a giant river. I haven't been there, but our family has kinsmen at Tillamook, so I heard about the white men there."

"Tell me what you heard."

"Well," she began, "the white men's canoes came down the giant river I spoke of before. That was only a short time ago, maybe two summers. There were many of them, white men and Indian people, too. No one knows where they came from, but the people say it was far, far to the east. There was a woman with them, too, a Shoshonee, I believe, with a baby on her back. They called her Sakagawea."

"And the white men, did they have names?"

She thought a while and then said, "One was called something like Cap'n Clark. He was the headman, I was told. The other one who was of importance was called Lewis."

"What did they want? What did they do?"

"They didn't seem to want anything much except for salt."

"Salt?"

"Yes, they worked very hard to get salt. They built a big fire and put sea water in pots and boiled the water away and kept the salt. They did that for days and days."

"Did they eat salt?" Cha it zit was amazed.

"They put it on all the food they ate.

"What sort of food?"

"As I said, everything. They ate fish, deer meat, bear meat, porcupine; the same foods we eat, only they cooked it differently."

"How?"

"They had big, hard pots, not of wood or basket. They could put the pot right over a fire and it wouldn't burn. They never used hot rocks, they didn't have to."

"Have you seen any pots like that?"

"Oh yes, the Makahs have them. They traded for them. They are wonderful. All the women at S'klallum wish for them."

"Amazing," Cha it zit said.

"They built a village and stayed a whole winter in it," she continued. "The men made all sorts of things. They had fireplaces to cook on that sent the smoke up through a special wall. No one had smoke in his eyes."

"They must be very wise."

"Yes, very. They can put their ideas on a white sheet they call paper and then go back and get the same thoughts later. They have many strange possessions."

"Do they have *spirit power*?"

"Yes, it is very strong, I am told. The man they call Lewis spoke of it often. They call their spirit power 'God' and he has a son they call 'Jesus.'"

"What did the Tillamook people think of the white man's spirit power?"

"They were impressed. Some of the people asked to be told more about it so they could get some for themselves. But this man, Lewis, said that he couldn't tell anymore, but that maybe a special man, a missionary they called him, would come later."

"Where are the white men now?"

"They have gone. They went back up the great river."

"And the people at Tillamook – did the white men change them?"

"In some ways, I suppose. They are talking of sending a group of people up the river to track Cap'n Clark and Lewis to see if they can bring back a missionary who will give them the white man's power."

"I can hardly believe that such things are happening," Cha it zit said quietly.

They were now nearing the treacherous rocks off the southern point of Samamao. Cha it zit jabbed his paddle hard and steadied the canoe while Tsexad resumed her position. They could see the canoes ahead slewing in the dangerous water. Then the current eased and the long channel that led to Swetquem was in view. It lay calm, sparkling in the summer sunlight.

At last they were home.

The early arrivals were already stowing away their equipment and food stuffs. Cha it zit, assisted by Tsexad, carried bales and boxes into his small bachelor compartment. She followed him in and waited while he placed the goods. He turned to find her there. Surprised, he began to speak, to send her away, but then he stopped himself. Clearly a decision must be made. Did she, as his slave, share his space? Or did she, as a slave, sleep where the other slaves slept, by the drafty door. Before she had slept near her mother, not as a slave, but as a daughter. What a problem. He solved it by ignoring it. And so it was that Tsexad chose and came to share Cha it zit's bed.

That night, when he lay down on his fur robe to sleep, he recalled how Xeweda had seduced him. "As I have shown you, show my daughter," she had said. He pulled Tsexad gently beside him. She did not resist, but he could feel her young body tremble. He stroked her until she relaxed. An exquisite tenderness came upon him. Never had he felt such warmth, such softness. He wondered then who was the master and who was the slave. After they had finished, they slept, the bearskin covering them both.

When they emerged the next morning, Cha wentz was standing nearby. He looked at Cha it zit in a sneering manner and said, "I see that my brother has taken a woman, a slave."

"She is only half-slave, she has noble blood," Cha it zit hissed at him.

"Ah, little brother is sensitive," Cha wentz laughed as he sauntered away.

Cha it zit ran to the beach to bathe. There were many things to be done, the first of which was to consult with Sax ump ki. During the food-gathering season he had added his catch to his father's, which was only right. But now, considering his new status, his manhood, he felt he deserved to keep the products of his labor. How else could he accumulate wealth to wager on sla hal?

He found his father on the beach deep in a discussion with fishermen from the other houses. They sat on their haunches, a few wore shredded cedar bark capes to ward off the morning chill, but most were naked, as was common in the summer.

"Cha it zit," Sax ump ki greeted him, "sit with us. There has been much talk of your fishing success. You are needed to fish with us at the weir on the Nooksack River.

Cha it zit was pleased. It seemed that each new day brought more recognition of his abilities. He sat, trying to appear nonchalant, matter-of-fact, when he responded. "Agreed. When do we go?"

"Tomorrow. There will be repairs to make – the river struggles against the posts each year. If you possess tools; a maul, ax, or knife, bring them."

"It should be a good season for the humpbacked fish," an elder, whose face was lined by many summers spent netting, spearing, trolling, and trapping for salmon, spoke knowingly.

"Yes," a younger, heavily muscled fisherman agreed. "I believe that many of this last fish tribe will come, so many that they will clog the waters."

"We had better ready our tools, load our river canoes, so we can get an early start tomorrow," Sax ump ki said.

The men rose stiffly and went their ways. Sax ump ki turned to Cha it zit. "You may already know how we trap the fish. Perhaps someone has explained it to you."

"No one. I only recently gained my fishing power. It came to me of its own accord."

"That often happens," Sax ump ki nodded. Then he continued. "The weir, which is placed across the main fork of the river, belongs to many of us. We built it together, and we fish together. Then we divide our catch."

"That seems fair, but does everyone in the villages own it?"

"Actually, it is the headmen and their families, those who built it. It is customary for us to fish at night when most of the fish come. Those fish we keep. Other people can fish during the day along with us. Those fish are divided among everyone. The fish come to give us food. It is a gift, so we all share. Everyone gets fish from the trap. That is the way it is."

"I see," Cha it zit said thoughtfully. "I came to ask that I be permitted to keep the fish that I bring in, but now I see that I will share in what everyone gets."

"You have a reason to want to keep what you take?"

"I must begin to collect wealth. The gambling season will begin. I'll need goods to wager in sla hal."

"Well said," Sax ump ki smiled. "My son is thinking of his future. Keep what you catch except at the weir. That must be shared."

The next morning the fishermen, using their shovel-nosed river canoes, paddled along the shore to where a sizable stream emptied into the bay. The water was shallow, filled in where the flow had slackened and caused the silt to settle. There the men used stout poles and propelled the shallow canoes against the current up the river. When the water deepened, the men used their paddles to propel the boats.

The land was level, so the stream was placid. It seemed to be scarcely moving between the low banks on either side. There was evidence of frequent flooding along the banks. Willow branched into thickets, and beyond them one could see cattails holding their brown tassels high on spikes. One could see small, low-lying ponds, the result of beaver dams, dotting the land beyond the small river. Red-winged blackbirds flitted among the reeds and overhead a red-tailed hawk circled aimlessly.

Cha it zit rode in Sax ump ki's canoe, helping to paddle up the sluggish stream. At his feet was a maul that he had borrowed from Sta nek. An ax and a knife with a jade blade encased in leather lay at his side. He was at peace. Time went by; no one spoke. There was no need for talking. Then a sound became audible, a half-whisper, half-murmur.

"We are nearing the main river," Sax ump ki said. "You are hearing the song it sings. All the land to the west of us and north to the main river, the Nooksack, is called Chachosen by our people. It is an island of sorts, with the sea water at the outer edges. It is the place of our permanent homes."

"Where is the weir?"

"On the main river, just past the forks."

"What is to the north beyond the place where the two rivers come together?" Cha it zit wanted to know.

"The river is blocked by a great jam of logs. The waters run under and around them in great whirlpools. The river is in agony there and the noise it makes is terrible. Some of the Skalakhans, the people we displaced, live there and the Nooksack people live beyond them."

"Are they friendly?"

"We don't war with them. They have their area and we have ours, but we don't associate with them, either."

"Why not?"

"They are different from us. We don't speak the same way. It is hard to understand them. They go into the mountains and the forests to hunt and they fish in a lake behind the hills. We go to the islands and live on beaches. We are different," Sax ump ki said, waving his hands.

"Don't we go to the mountains, too? I have heard stories of such trips."

"Only those of us with great spirit power. There are beings in the forests and high places with the ability to harm us."

The noise of the river grew, and beyond a bend they saw it swirling and rolling on its way to the sea. Expertly, Sax ump ki turned the clumsy dugout into the current and they were carried along toward the west. Now it was Sax ump ki's task to guide the shovel-nosed canoe along the river's edge. Ahead they could see the upright posts of the weir.

The others had landed and were examining the fish trap.

"The river has changed course, we have to redo one side of the trap," one observed.

"These posts are weakened," announced another, who was standing on a walkway above the weir.

"No use worrying about it," Sax ump ki said, as he hoisted himself from the shovel-nose, which he had run up onto the river bank. "We'll need to cut more young fir trees, and gather cedar branches to make ropes for tying."

Before taking his ax in hand, Cha it zit studied the weir. He admired its plan of construction. A series of poles placed in pairs a short distance apart and tied together at the top had been positioned across the river, reaching from bank to bank. Long, slender poles had been laid in the crotches of the twin posts and tied in place by stripped cedar branches. In addition, two rows of parallel poles had been tied to the upstream side of the slanting poles, one above and one just visible below the swirling water. A webbing had been constructed of willow limbs and tied in place to the parallel poles on the upstream side of the slanting posts. An opening about the length of a man's arm was left in the center. To that, a door made of looped cedar limbs was attached, so it could be opened and closed. Upstream from that was another frame with webbing on it and with sides that extended downstream to intercept with the trap, forming a holding pen. A platform was built alongside the pocket for the netters to stand on.

"Very clever," Cha it zit thought as he took his hatchet and followed the others to a stand of young trees. He had known of the fish trap but this was his first look at it. "Yes, very clever," he admitted with a sense of pride, proud of the fact he was to be part of the fishing party.

The work was hard, cutting trees, splitting cedar branches, weaving willow, tying, digging in new posts, but the men bent their backs to it and the weir was ready for the time when the fish came.

"It will be a few days yet," Sax ump ki told Cha it zit. "When our lookouts see the salmon start up the river, we will return. For now, let us continue down the Nooksack toward the salt water and see how the fortification at Temxwigqsan is progressing. It should be finished."

"Or nearly so," Cha it zit commented.

The dugout moved effortlessly on the sluggish current. The river spread itself into a delta. Channels oozed lazily toward the inlet where they consorted with the clear green-blue waters of the strait. In the distance, the familiar hump of Samamao rose out of the sea. The sea was calm, blocked to the north by a point running far out, forming a sharp angle to the shore.

"We call that point Steyuqsun," Sax ump ki said as he nosed the dugout south along the mainland. They passed a slough, the other end of the one that emptied into the small bay near Swetquem. The water was brackish, fed from both ends by tidal flow.

"When the moon is at its highest power," Sax ump ki said, "then the water runs full through here. It separates our villages, Momli, Temxwigqsan, and Swetquem from the others, then we are on an island. At one time, it appears to me, it was a true island."

Cattails and bulrushes filled the slough, and herons waded among the rushes, poking into the ooze for small fish, shore crabs, or whatever foods they could scavenge. They took flight, rising awkwardly, their long legs trailing behind them. Their raucous cries, throaty and hoarse, echoed across the water.

"They are valued for that sound," Sax ump ki said. "Our people call them the village guardians. It is hard to get past them without alerting them. Then they call out, 'Someone is approaching.' "

"I can see that," Cha it zit laughed.

"Never kill a heron," Sax ump ki added, his face serious. "They are our friends."

The longhouses of Momli lay ahead, situated in a small cove.

"We'll stop here to tell them that the weir is ready. All of the Lummi people have the right to dip fish from it, whether they own it or not."

A few elders came to the beach to greet Sax ump ki and his son. "The able-bodied villagers have not yet returned from food-gathering," they explained.

Among them were several ancients from the Semiahmoo people, who inhabited the land situated north of the Nooksack delta.

"Come and drink nettle tea with us," they invited.

"The rest will do us good," Sax ump ki said, pulling the dugout high on the beach away from the clutching fingers of the incoming tide.

They sat in the dark longhouse, a smoky fire adding to the soot that had accumulated around the smoke hole. The tea was hearty and Cha it zit felt it relaxing him – warming his insides, soothing his soul.

"Nettle tea is good for stomach hurts," an old crone was saying as she offered refills from a mountain goat horn dipper. Cha it zit was contented.

"It is good," he thought, "now that I am a man."

One of the elders, a grizzled, shrunken man with the eyes of an eagle, suddenly spoke. "A strange thing happened a day's journey north of us where the great river flows into the sea. That is where the Musquem people have their villages. I know that what I am about to tell you is true, for my daughter is married to a Musquem and I was there at the time."

He stopped speaking to emphasize the significance of what he was about to say. The others waited patiently while he sipped his tea and savored his newfound importance. At last he continued.

"I was at the village, as I said, when many canoes came down the great river. They were not like any canoes I have ever seen, but it was those in the canoes who were so remarkable. They had hair, long shaggy hair, half covering their faces. What we could see of them was skin of a pale color, and strange eyes. Some eyes were as blue as the summer sky, some were the color of the sea on a stormy day, others were like wet sand. Not one of them had eyes of the proper shape and color," he repeated.

"You should have been there to see the hair that sprouted on their heads. Some of it was the color of dried grass, which indeed we first thought it was. Some was the color of the red clay we make face paint from, some was like the yolks of bird's eggs. There was hair that grew in spirals and hair that stood out like a bush. They were strange to see."

He sipped his tea, regarding the listeners solemnly and then spoke again.

"I can hardly describe their clothes, except to say that almost all of their bodies were covered in blanket-like material." He looked down on his own nearly naked body and then added, "I do not see how they could move in all those garments."

Cha it zit was fascinated. Here was another piece of information about the aliens, as he had come to think of the white men. This time they appeared coming down a river that led to the inland country. Were they everywhere?

"They made strange noises which we could not understand," the old man was saying, "but there was one, of our appearance, who spoke in a tongue similar to ours. They spoke to us through him."

"What did they say?" Cha it zit was so eager to learn all he could that he forgot protocol, that he should not interrupt, particularly elders who were allowed to speak as long and as slowly as they wished. The old man gave him a disdainful look, and seemed to deliberately prolonging the tale.

"The headman told his name," the elder said eventually. "Simon Fraser, I think he said. It was hard to pick out the words. He said they had come from many many day's canoe trip away, across many rivers. He asked if we had furs. We told him we had a few beaver and sea otter, but that we spent our time fishing, not hunting. This Simon Fraser seemed unhappy to hear that. Then he asked if there were many others like us living on the river or by the sea. We told him not all the people were friendly like we are and to beware of the Cowichans across the water, for they are very fierce. He asked if they hunted for furs, and we told him that they fished mostly, too."

Again the old man stopped, only this time he took out his pipe and filled it. After a few puffs, he continued.

"Then that man and all those with him got back in their canoes, and followed the river down to the gulf where it enters the big waters. They tasted the water and spit it out. Then they went back up the river."

The old man spread his hands in a gesture of bewilderment and shrugged his bony shoulders.

"Is that all?" Cha it zit asked.

"That is what happened."

The group was silent for a while, each considering what he had heard. The old man puffed on his pipe, sending a trail of knik knick smoke to blend with the wisps from the fire. He was pleased with himself. But then he laid down the pipe and spoke again, a new thought having come to him.

"That is not the first time that I have seen those strange beings. It was at the fishing grounds across the water from our village, Semiahmoo, where a large canoe with wings of white came upon us. We were reef-netting for salmon as we do each summer. They came so quietly that they were almost on our canoes before they called to us. We couldn't understand them and they couldn't understand us, but we used gestures and told them that while we didn't have many furs, we traded with others who did. Then they went away."

"They always want furs," Cha it zit commented. "Why do they want furs, especially sea otter fur? They can't eat it. They seem to have too many robes already. I don't understand it."

The old man shrugged again. "Maybe they are not men. Maybe they are spirits. Maybe they need fur to survive. Who knows?"

"It is strange," the listeners agreed.

"We are rested and the day is almost past, we must leave for Temxwigqsan," Sax ump ki announced, rising.

They followed the shore, now curving so that they faced toward Samamao, which was bathed in late afternoon sunlight. They passed a fishing camp known as Telapi. There were people there, women preparing salmon for drying and men lounging about their fishing canoes. Sax ump ki called to them, telling them that the weir was ready to use as soon as the humpbacked salmon chose to go up the river. They reached Temxwigqsan as the sun was slanting low rays across the sky.

"There it is," Sax ump ki shouted, pointing with his paddle. "How strong it looks." The palisade was in place, completed around the village. It stood eight feet high, the poles reinforced by top pieces as Sax ump ki had requested.

"That has been my dream," Sax ump ki exulted.

Cha it zit had never seen his father so pleased before. He looked at him in surprise, this strong dominant man,

this accepted leader, was like a child in his pleasure. His smile, showing strong teeth, was wide, and his eyes shone.

"Ever since that terrible time when many Lummis, including my wives and children, were killed by Yakultas on that accursed island, I have dreamed of a defense."

"And now we have one," Cha it zit remarked.

"Yes, now we have one, son. From the barricade we will not only defend ourselves, we will launch our own attacks. The Yakultas, and any other enemy who dares maraud against us, will feel our war clubs."

"Your war clubs," Cha it zit boldly corrected him. "I am not a warrior."

Sax ump ki was too elated to notice or to react to his younger son's statement.

They had reached Temxwigqsan. Several slaves came down to the beach to assist in bringing the dugout to the bank and secure it. Xecusem, as headman, met them at the entrance of his part of the double longhouse.

"I see that you have your fine young son with you," he said by way of greeting to Sax ump ki. Cha it zit sensed an eagerness on the part of Xecusem. There was a look akin to greed on his puffy face.

When they had been seated at the fire of his first wife and had been offered a bracing mint tea, Sax ump ki commented on the defense wall. "Your oldest son, Cha wentz, is a born leader," Xecusem said, eyeing Cha it zit. "He made the warriors work hard. Those two S'klallums did well, too. You have two fine sons, my friend."

"Three," Sax ump ki said, pride in his voice.

"Three?" Xecusem was confused.

"My wife, Wanana presented me with a son."

"That is good. All I have is daughters." He looked directly at Cha it zit. "I see that this son has changed since last I saw him. See how strong he has become." He laughed. "And tall. He looked me directly in the eyes. Yes, I would say that this one has become a man. Eh?" He poked Cha it zit in the ribs with a fat finger. Then he turned directly to Sax ump ki and said, "You have not forgotten what we spoke of?"

"No," Sax ump ki answered, "but that is not why we come at this time."

"Oh?" There was dismay in Xecusem's voice. "If not that, then what?"

"The weir has been repaired. When the last salmon tribe runs up the river, you and your people may dip fish."

"That is good. But about this other matter. There will be considerable goods to accompany my oldest daughter. The man who gets her as his wife will be fortunate. She has been well trained in women's arts, she is strong, she will bear many children. The alliance will be good for both of our families."

"I know that," Sax ump ki answered. "I have spoken of it to my son here. He is in agreement."

"He is? Then it is settled?"

"No, there are formalities." Sax ump ki resisted, not wanting to be pushed into anything.

"I would save your son the ordeal in store for him."

"The ordeal?" Cha it zit entered the discussion even though custom decreed that he remain a bystander.

"A suitor must wait outside the house of the girl's parents until he is invited in or is so ignored and abused that he leaves. This could take days. Then, if he gets inside, he must still wait until the people pity him and make him welcome or, again, he leaves. If he is accepted and fed, he will become the girl's husband," Sax ump ki explained.

"Then," Cha it zit answered, "I will spare myself the ordeal, since Xecusem seems to feel that I will be a suitable husband for his daughter, I ask now that she becomes my wife."

Xecusem's eyes widened. He broke into a grin. He laughed and slapped Cha it zit on the back. "Well spoken," he said. "You will see. My daughter will make you a fine first wife."

"There are some things I will ask before the commitment is made," Cha it zit said, his face serious, his voice trembling with ill-contained emotion.

"This is unusual. It is against custom," Xecusem objected.

"That is true," Sax ump ki agreed. Turning to Cha it zit, he said in a hard voice, "I and my friend here are the ones to settle what is to be done. We make the arrangements, you have nothing to say in this matter. This is a family thing. It has to do with inheritances, names, privileges, fishing rights. It is bigger than my son or his daughter. Can you understand that?"

Cha it zit felt abased, ashamed, but his goal, his dream, over-rode everything. He controlled his feelings and answered in a flat tone, "Of course. I agree that there is much to consider. Great wealth and prestige will change places. But hear what I have to say."

"We can at least listen, but it is very unusual," Sax ump ki relented.

"I want the right to build a house here . . . not only a house in which to live, but a house in which to hold potlatches, feasts, and sla hal games."

Xecusem laughed. "You, with nothing but your family behind you, talk of building a house? You talk of potlatching?"

Cha it zit stiffened with determination. "I have a gambling spirit. You have seen that. I have other powers you know nothing of. They need space. Maybe the longhouse will not be built now, but it will come to be. I want that right." Then he changed his tone, becoming conciliatory. "You should be happy. Your daughter will be close. You will see your grandchildren grow up."

"Yes," Xecusem said in a kinder voice. "I grant you that, and the help you will need to build a house."

"Then it is settled," Cha it zit said, speaking firmly.

"When the spirits come during the time of the rains, we will meet and speak of this again," Sax ump ki announced, rising.

Cha it zit was stunned. Without forethought, he had committed himself, with a few words, to a lifetime of association with a girl-woman he had never spoken to.

All the way back to Swetquem, through the cut, Shalahsheen, along the shoreline of the deep bay, he thought of his future. And always the face of Tsexad, his slave, his friend, his lover, was before him. How would this marriage affect her? Would it change their relationship? How could he tell her?

Then, without him willing it, a vision of the Samish girl appeared. Her eyes were soft, as he remembered her, but now they were sad, too. He forced her from his mind. "There is no place for her," he realized.

12

The River

Cha it zit felt that he was being carried on a current, tossed randomly from place to place without the least ability to interfere with or alter the course of events. So much that had happened was beyond his control. It seemed to him that spirit powers were playing with him, propelling him toward some mysterious destiny. He had to think, he had to try to make some sense of the events of the past days.

Looking about him, he saw that the family was occupied with immediate needs. Fishermen were readying their dip nets for working the fish trap, hunters were preparing bows for the annual autumn trek up the river into the high country for deer, elk, bear, and mountain goat, while the women, always busy, were going quietly about their home tasks. "No one will miss me," he thought as he slipped out and climbed toward the rocky perch that had been his sanctuary since he was a child.

There, with the sea and the islands laid out below him, he sat and let his mind wander. "All that I can see is Lummi domain," he thought. "Once I only saw it, now I know it. Some day my own house will stand where I can look across to Samamao and beyond to where Allulung lies in ruins."

He thought then of his impending marriage to a female person he had only seen from a distance. What would she be like? A picture of Klatok came into his mind. Would his wife be like Sax ump ki's first wife, demanding, powerful in her own right, a leader of women? Even Sax ump ki sometimes cowered before Klatok. Cha it zit dismissed that image and replaced it with one of Tsexad. He saw her, soft, pliant, wishing to please. She could never be a wife, especially not the first one, he knew, but he wondered what his marriage would do to their relationship.

"It won't change," he swore to himself. "This marriage is only an alliance for status, it won't change anything." But even as he vowed this, he knew in his inner mind that it couldn't be so.

Then he reflected on his newfound prestige, on the powers he had displayed quite beyond his control. It disturbed him that the shaman spirit, the one that his dead great-uncle Tselique had insisted he would have, manifested itself so strongly at the time of his brother's birth. "Not that I'm not glad I helped him come," he admitted, "but I have no wish to be a shaman – and yet I seem to be one."

He knew that the people now treated him with respect, even awe, and though it pleased him, he missed the easy relationships he had known before. "How can I suppress the spirit?" he asked himself.

That was when he heard the raven light on a branch above him, his wings rustling, as only a raven's do. "There is no way," he said aloud, looking at the large black bird. The raven cawed twice and then flapped away to light at the top of a giant fir some distance off. Cha it zit called to him, "Caw, caw," in the raven tongue, and it answered.

He sat, letting the thoughts come as they would. Below him, a distance from the shore, a silver streak appeared in the water. It glinted in the sunlight of the morning. "The humpback fish have come." He jumped up and ran toward the village. "The fish are coming," he called out.

People popped out of the low doors. "It is time," fishermen called to each other.

Sax ump ki took charge. "The salmon people of this tribe have decided to travel up the small river to the mountains. Others will be coming in from their villages under the sea to journey up the main river where they will enter our trap and give themselves to us. We must hurry to meet them there. Let the women come when they have made preparations."

Soon the small river which emptied into the bay near Swetquem was filled with people poling or paddling river canoes upstream. Cha it zit was eager to take his place on the platform above the trap. If he truly had found a fishing spirit, this would be the time to prove it to himself and the villagers.

When he arrived, he found that others had already stationed themselves at the trap, nets in hand, waiting for the first salmon of the run to appear. Suddenly there was a shout, the waters began to seethe, and the trap filled with silver fish. The men sprang into action, netting the struggling fish and tossing them onto the banks. Once the run of the fish began, the salmon were driven by the urgent need to spawn, and there was no respite for the netters. The salmon arrived in such numbers that they threatened to choke the river. The trap remained filled no matter how fast the men worked. When it became completely glutted, the opening into it would be closed until it had been emptied. Cha it zit, with others, mostly boys, retrieved the fish that had been tossed onto the land and piled them in the shade to keep cool until the women arrived to begin to dry them.

One of the netters called to him, "Take my place," and Cha it zit, raced out onto the shaky platform. The river ran clear, blue-green water flowing over a sandy bed, its current rocking the boards on which the fishermen stood. The netter handed his long-poled net of nettle twine to Cha it zit with a look of relief and departed.

Cha it zit held the heavy net awkwardly. It wasn't as easy as he had thought to dip it, pull it up full of fish, and then toss the fish, all in one long, fluid motion.

"Your first try?" a man next to him asked.

"Yes," Cha it zit admitted, embarrassed.

"Do it this way," the fisherman demonstrated, moving slowly.

"I'll try." It took some practice, but soon he was working with an easy rhythm, keeping up the with others. But after a while his body began to ache. He forced himself to keep working. One by one, the netters signalled for relief and others took their places, but Cha it zit kept doggedly netting and tossing.

The women arrived and began to butcher the fish. Even the elders came to do what they could.

The run lasted for several weeks before it slacked off. The fish that came then were stragglers following in the wake of the more vigorous leaders.

"It is time to open the trap," Sax ump ki said. "Let the rest travel to their destination. We have plenty."

The run was a record one. It would take many trips by dugout to transport the cured salmon to the villages. The people were satisfied. There would be no hunger during the coming winter.

As he had been promised, Cha it zit received a share of the dried, smoked salmon. Tsexad placed it in new storage baskets she had made and put them on a rafter above Cha it zit's bed. Then she took out some basketry materials and began to work on a partly finished basket. It was of a strange design, beautifully made.

Cha it zit lay on his bearskin robe watching her with interest.

"It is a *twana basket,*" she said. "I am making it for you. See the design?"

"It looks like a tree."

"No. How can you say that? It is a hand. See, here are the fingers."

"What does it mean?" he asked.

"It doesn't have to mean anything. But do you like it?"

"It's nice," he admitted. "But why are you making a basket for me? Whatever a woman makes belongs to her. That is our way."

"I know. But I heard that you will be taking a wife," her voice became wistful. "You will be needing gifts to give for her."

"Who told you that?"

"Everyone knows. It is true, isn't it?"

"Yes," he admitted, "it is true. But that won't change things between us."

"No," she agreed, bitterness in her voice. "I will still be your slave and you will be my master."

"I don't think of you that way," he told her. "I can't help what your mother did, running off with a northern slave, or giving you to me."

She gave him a look of hurt surprise, threw down the basket, and ran from the alcove.

Cha it zit started to follow her to explain his own feelings about the marriage, but then he checked himself. How would it look? How could he chase after a slave girl? How could he debase himself before her? He was confused. His feelings of regret for saying what he did and his sense of pride fought, neither winning. Whenever he had a problem, in the past, he had spoken to Latsi. Suddenly he wanted to see her. Even if it was considered weak, beneath a man, to seek advice from a woman, he felt a great need.

Latsi was nowhere. He spent most of the day nosing about the places where she might be. Then he saw Xedewa.

"Have you seen Latsi? " he asked, knowing that Xedewa would not laugh at him or accuse him of weakness.

"Yes," she admitted, "but it will be many days before you will see her. She has received the sign that she is becoming a woman. They have secluded her in a small house."

"Where?"

"You must not go near her. She now possesses power to harm you. Only old Tseutsi attends her."

"My sister is a woman?" Cha it zit was incredulous. "I have been so involved in my own manhood that I forgot that she is growing into maturity, too. I have lost, her haven't I?"

"Perhaps," Xedewa answered. "When her period of isolation is over, when Tseutsi has taught her all she should know to be a full woman, to be a wife, suitors will come. Then she will be given to some man and will go. But you, I know that you are to marry and bring your wife to this house."

"You heard?"

"I heard."

He could not bring himself to talk to her about his emotions or about how he had hurt Tsexad. Instead he asked, "What will happen to Latsi in the hut?"

Xedewa studied Cha it zit's face, her look so intent that he shifted his gaze to avoid meeting her eyes. "It is difficult to speak of such things to a young man," she said, "even when it is his own sister who is being discussed."

"I would think that you could tell me anything," he replied, "we are bonded by the experience we shared. You brought me into manhood, cannot you tell me of Latsi's entry into womanhood?"

Xedewa's flint-hard eyes softened. "I perceive that you care for your sister and also that I can speak openly to you about things not usually discussed."

Cha it zit nodded as they spoke. They had wandered aimlessly, and soon had circled to the back of the longhouse and were nearing the small creek which flowed amid a tangle of brush toward the bay. The secret place where he had played as a child, the same one he had used as a hiding place when the Yakultas threatened, was nearby.

"Come," he said, "if this is to be a confidential talk, let us have it in privacy." He guided her through the opening in the bushes, now grown over, and into the small, cleared space. "Now, tell me about Latsi's ordeal."

"It is hard to endure the requirements put upon girls when their first unclean period comes. It is even harder for high-caste girls like Latsi. First, there is the pain that may come, and the fear. But that is nothing compared with the ordeal itself. As soon as there is a showing of blood, she is removed from any contact with people."

"Why is that?"

"Because she then possesses magical powers so dangerous and potent that she cannot control them. She can cause terrible injury. She can even cause men to lose their spirit power if they come near her. If she were to go near a salmon stream, the salmon would never come there again. Berries would wither on the vine if she passed by."

"I can see that girls in that state are to be feared," Cha it zit conceded.

"Yes, they must be confined for their own safety as well as for that of others. Besides, they are open, then, to receive *spirit gifts.* Girls must not eat or drink during their seclusion. They do not wear clothes, they must not touch their own bodies. If they need to scratch, they must use a head scratcher. There are many restrictions. It is important that they keep themselves very clean."

"How long must they endure those conditions?"

"Different clans have different rules. My people, the S'klallums, are rather lenient. The time is six to fourteen days. I believe that the Snohomish people keep a girl in absolute seclusion for twenty-six days. I do not know what your people require."

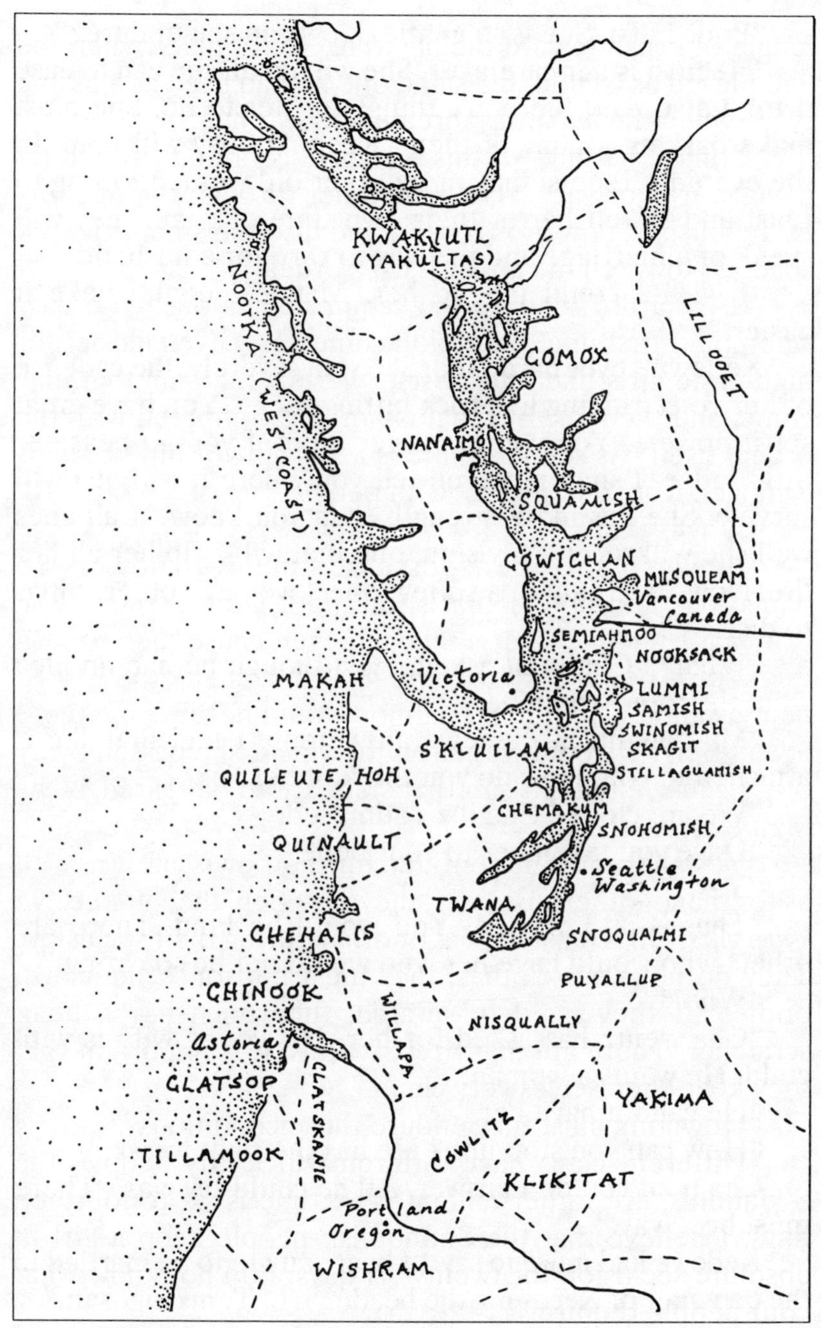
KWAKIUTL
(YAKULTAS)
NOOTKA (WEST COAST)
LILLOOET
COMOX
NANAIMO
SQUAMISH
COWICHAN
MUSQUEAM
Vancouver
Canada
SEMIAHMOO
NOOKSACK
MAKAH
Victoria
LUMMI
SAMISH
SWINOMISH
SKAGIT
S'KLULLAM
QUILEUTE, HOH
STILLAGUAMISH
CHEMAKUM
SNOHOMISH
QUINAULT
Seattle
Washington
TWANA
CHEHALIS
SNOQUALMI
PUYALLUP
CHINOOK
WILLAPA
NISQUALLY
Astoria
CLATSOP
CLATSKANIE
YAKIMA
COWLITZ
TILLAMOOK
KLIKITAT
Portland
Oregon
WISHRAM

"Poor Latsi. She is so gentle. How can she endure?"

"Tseutsi is her caretaker. She will do all she can to ease it for Latsi. And there are things for her to do. She must make baskets – many of them. Keeping busy will help. In the evening, Tseutsi may bring other old women to sing to Latsi and to tell her of their own time of trial. They will speak of a marriage and the way to care for a husband."

"I wish I could talk to her. Maybe I could make it easier."

Xedewa's eyes hardened. She said severely, the cadence of her voice ringing like rock hitting rock, "You have great spirit powers. You will endanger them if you go near her little lodge. I shall not even tell you where it is. Latsi will survive. She has more strength than you know. If all goes well she will receive a vision, one that will help her all her life. Now," she said in a softer tone, "we have other things to discuss."

"What?" Cha it zit asked, even though he had an idea what the topic would be.

"My daughter," Xedewa answered. "I feel that she is attached to you. How do you think of her?'

"We are close," Cha it zit admitted.

"Have you taken her to your bed?"

"Yes."

"That is as I wished. You would be kind. There are others who would have her who would not be so caring."

"Who."

"Cha wentz has asked for her. His young wife is with child. He wants a woman."

"He cannot have her."

"How can you stop him? He has the right to ask."

Cha it zit couldn't answer. All he could say was, "There must be a way."

Xedewa had more to say. "I hear you are to be married to the daughter of Xecusem, the headman of Temxwigqsan."

"That is true."

"Have you seen her?"

"Only from a distance, but her appearance isn't important to me."

"I know," she said with an edge to her voice. "But her father's prestigious position is. And the village, you like that location, don't you?"

"It was his idea. He and my father have sealed the arrangement."

"And where will Tsexed fit into this?"

"There will be no change. If I could have made her my wife, I would have."

"There may be a way."

Cha it zit thought, "She has to come to the point of all of this at last." He asked, "And what is that way?"

"If a potlatch were given for her, one honoring her, a naming perhaps, and if sufficient gifts were given, her status could be raised and she could be acceptable as a second wife."

"Yes," Cha it zit agreed, thoughtfully.

"I have been asked by many to give them medicines; they pay me well. In time, I can accumulate much wealth." She was speaking rapidly now, encouraged by his apparent interest.

"I can help, too," he said, "I can win much at gambling."

They both sat in silence for a while. Then Cha it zit spoke, "But that doesn't change a thing. I am committed to marry and I will."

"Yes," Xedewa agreed. "That comes first."

The conversation was over, each had received answers to worrisome questions, and it would have been logical to leave that place, but the sun was warm, and the moss on which they sat was comfortable. So they remained in silence, pleasantly relaxed. At last Cha it zit said, "Tell me of your lover, Tsexad's father. What was he like, that Yakulta?"

"Arrogant, even as a slave; strong, proud."

"Was he kind to you?"

"Usually. He was kinder than my husband. I needed him to help me escape and to help me stay alive when we were free. Why do you ask?"

"Because the Yakultas are our enemies. If we are ever to defeat them, to keep them away from our villages, we must know their strengths and weaknesses."

"He spoke of battle often. To fight was his life. I can tell you this, there are others, Kwakiutl people, as the Yakultas are, who are more to be feared. And beyond them are the fierce Haidas. They have dealings with the white men who come from the north to trade for furs."

"There are aliens to the north?" Cha it zit was incredulous.

"Many of them. They are dangerous – there have been battles between them and the Kwakiutls. They have *thunder sticks* that they can point at a man and kill him. Now the warriors of the north have them, too."

"You know this to be true?"

"I have seen and heard them myself, at our village. I know they exist. I believe that it is true that our enemies have them, yes."

"The white aliens are coming from everywhere," Cha it zit observed with alarm. "They come from the great rivers to the north and to the south of us. They come from the sea to the north and south of us. From what you have told me, there is no place from which they cannot come. I fear that more and more will arrive. Our lands will not be ours, the rivers will not be ours, the sea will not be ours, the fish will not be ours."

"That may be true," Xedewa said, matter-of-factly.

"What can we do?"

"We cannot fight them," she replied. "Their power is greater than ours. We must get their power for ourselves."

Cha it zit rose to go. He had much to think about and much to do.

"I will stay a while longer," Xedewa told him. She too, had thoughts to think.

When Cha it zit returned to the village, he found What la cum standing on high ground in front of the longhouse. Across the water, beyond the foothills, the cone of a mountain named Kulshan stood snow-covered and bold against a graying sky. A "V" of geese, flying low, circled and landed in the waters of the bay.

What la cum turned as Cha it zit approached.

"I feel in my bones that the summer is nearly past," he said. "The elk and deer will be coming down from the high country. Already the geese are gathering in from the north. It is time to hunt."

"Will you go up the river?" Cha it zit asked.

"Yes. There are prairies up there," he pointed toward the mountain, "where the river begins, where the elk gather in herds. One can get beaver and deer and bear there, too."

"I am not a hunter," Cha it zit stated emphatically, "and I have no wish to get a hunting spirit, as you know, but I would like to see that meadow, and to follow the river to the mountain."

"I will speak to the others," What la cum offered. "Some of the women will go, too, to pick the blueberries that grow high in the hills. You could help pole the river canoes, and carry meat. You could be useful to us."

"I would work hard," Cha it zit promised.

13

The Hunt

Cha it zit searched for Tsexad. At first he was annoyed by her absence. He missed the attention she gave him. Then he became concerned. Even though he knew his thoughtless words had hurt her, he depended upon her loyalty and sense of duty, if not affection, to bring her back. Several days had passed and she still hadn't appeared. No one had seen her, or if they had, they wouldn't admit it. After all, what was the sense of worrying about one slave girl?

Then he thought of Cha wentz and his desire to take her for himself. Cha it zit then became angry. If his half-brother had harmed Tsexad, he swore, he would have revenge. Then he stopped his runaway thoughts. Revenge against Cha wentz? Cha wentz the fearless warrior? Cha wentz whose powers were far greater than his own? What insanity. Still the anger was there. Perhaps there was another way to avenge Tsexad. Wealth. He could use his gift for gambling and pauperize Cha wentz. Feeling a bit mollified, he continued his search for Tsexad.

Suddenly, unexpectedly, he stood face to face with Cha wentz. He noticed with surprise and delight that he not only could look the powerfully built warrior in the eyes, but that he also actually looked down on him. "He outweighs me," Cha it zit admitted silently, "but I am taller, my arms are longer. I can outreach him." Cha wentz no longer seemed as formidable as he once had.

For a moment neither brother spoke. Then it was Cha wentz who broke the silence. "I have something to ask of you," he said.

"What do you wish?" Cha it zit kept his voice even, though he secretly exulted. It seemed possible that Cha wentz had not yet taken Tsexad.

"My wife is with a child, I am barred from her bed."

Cha it zit interrupted, "You wish my slave Tsexad."

"That is customary. I don't know how she came to belong to you, but, as my brother you cannot deny me."

"She is not here," Cha it zit admitted honestly. "I thought she might be with you."

"I have not seen her, though I have been looking."

"Brother," Cha it zit said softly, "I tell you this. If you touch her, there will be bad blood between us. You may be able to kill, but I have powers, too. Let her be. To hurt each other would only bring harm to the family. Think of it. The family is more important than either of us."

Cha wentz stared at Cha it zit in surprise. "The frog has teeth," he muttered, turning his back and striding away.

Cha it zit grinned. "I am indeed a man," he exalted. He felt that he would burst with elation. "I turned Cha wentz aside."

He continued his search. "Tseutsi," he decided. "She knows everything that happens." When he found her, she was scurrying, like a small, thin-legged brown bird, on some mysterious errand. In her hand she carried basketry materials. When he asked her if she had seen Tsexad, she only smiled and said, "I have no time for your questions. Tsexad is yours, I have heard; if she chooses to hide from you, that is your affair." Then she hurried off.

Cha it zit was amazed. Never before had his old grandmother spoken to him like that. She knew something, he decided. He could follow her, spy on her, but that would be demeaning. He let her go and abandoned his search. Somewhere, he felt, Tsexad was safe. She would return when she was ready.

The hunters had readied their bows, arrows, spears, and knives. Their river canoes were loaded and ready. Those women who were to go were waiting at water's edge. Cha it zit watched, wondering if he was to be included. What la cum had not approached him and Cha it zit was too proud to ask a second time. So he merely stood there.

What la cum looked his way, shouted something to Sax ump ki, and then beckoned, "You can pole for Klatok." Cha it zit joined the group, grateful to be included. He noticed that Xedewa was going along with other wives of Sax ump ki and What la cum. Wanana and her small son would not go, nor would Cha wentz's wife. Cha it zit had hoped that Tsexad would be with her mother but she was not to be seen.

The flotilla of canoes, joined by others from the village, left amid shouts of good wishes from those left behind. The first part of the trip was familiar, going up the small arm of the river to where it departed from the main stream of the Nooksack. From that point on, it would be new territory for him.

Cha it zit took the long pole when the river ran shallow and pushed from the stern of the small shallow-draft craft. Klatok stood in the prow. She was an expert canoeist and she knew the river, every twist, turn, trick, and peculiarity of it. Cha it zit could feel the strength of her thrust as she added her force to his. Occasionally she called instructions to him, push this way or that, ease up, or lean hard on the pole.

They reached the main flow and shoved into the current, struggling upstream. The water deepened and they exchanged poles for paddles. There was little conversation among the people. Energy was conserved for the great effort of progressing up the river.

At one point, Klatok turned toward him and said, "This all belonged to people who called themselves Skalalats. That was before we came here from the islands and chased them back. We are coming to the territory of the Nooksack people. The same name as the river – 'People of the Ferns,' it means."

"Will they let us pass?"

"We cannot talk to them – their language is different – but we have an understanding. They can take clams on beaches we share, and we can use the river."

"That is good."

The scenery along the river was slowly changing. The growth was heavy, willow, maple, cottonwood and alder trees crowded the bank cutting off the view of the valley the river had made. The river serpentined through the flatland, making the canoeists travel several miles across for every mile forward. It became monotonous, paddling and poling up the confines of the river.

They rounded another of the many bends and were confronted by the sight of a tremendous log jam. It seemed to be endless, reaching from bank to bank. Gigantic trees, some with roots still intact, lay in grotesque heaps, one atop another. Cha it zit was dismayed.

"What can we do?" he called to Klatok.

"We get out and walk," she said with a shrug.

"Walk?"

"We carry our supplies, then we carry the canoe."

Others were landing in a small cove below the jam. Ferns grew in abundance beneath spreading maple trees. It was an inviting spot, one where an afternoon could be drowsed away. The people rested there for a while, eating dried berry cakes and lying in the shade. Then, at a signal from Sax ump ki, they began to unload the canoes. The women were laden first and they began the trek to the end of the dam on a well-worn trail. Then the men toted the dugouts on up the path a mile or so to where the Nooksack

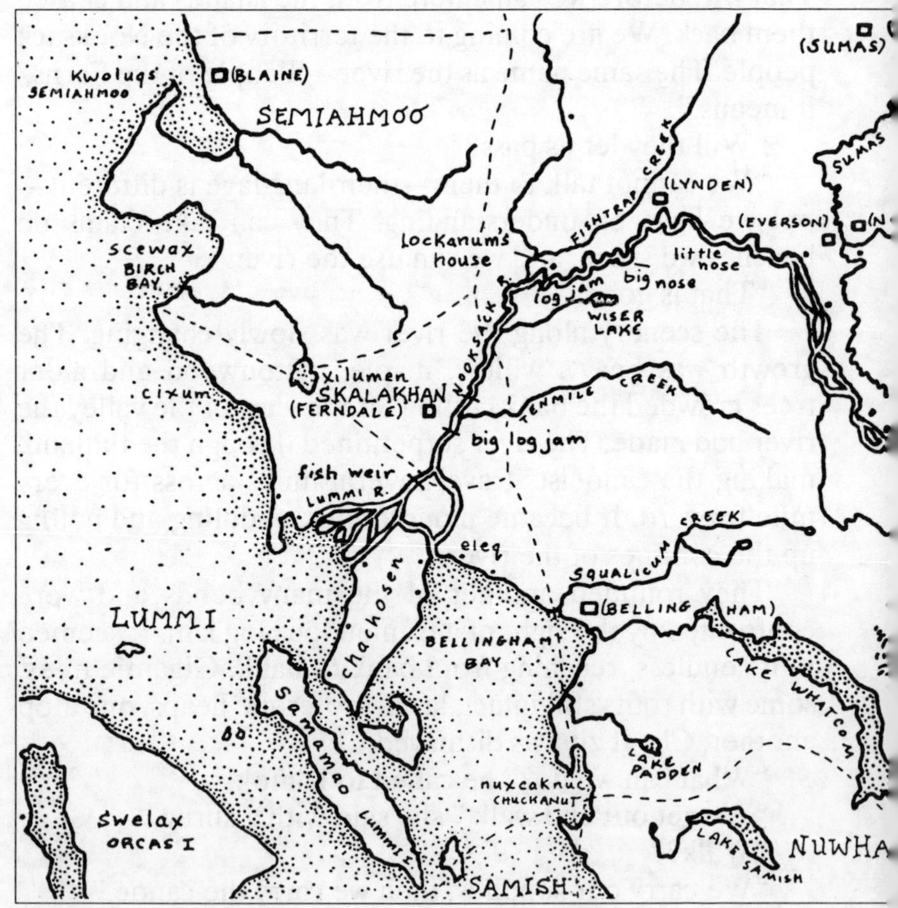

ran free above the dam. It was heavy work, carrying the dugouts, and Cha it zit marveled at the strength of the men as they struggled along.

Cha it zit noticed that large trees were growing in the jam itself. "It has been building up for a long, long time," he realized.

The scene was desolate. It spoke of death. It spoke of tragic waste. The river itself, so full of vitality and life, seemed to have died somewhere beneath it. It spoke, too,

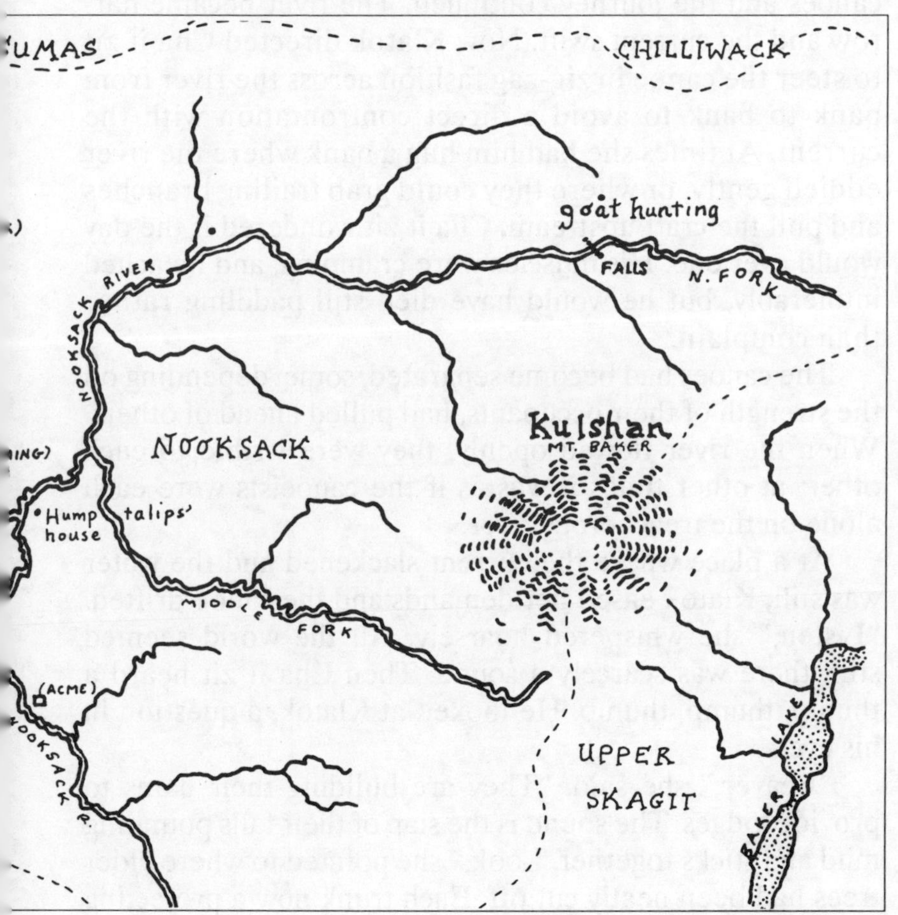

of forces of destruction so great that it seemed man could have no impact on them. Cha it zit found himself hating the dam as much as he hated the intolerable weight of the canoe he was carrying.

After what seemed to be forever, they came to the beginning of the jam, or the end of the river, Cha it zit wasn't sure which. At that point, the water simply slid under the logs. There was no protest; it seemed to accept its fate and disappeared.

The women were waiting. They helped reload the canoes and the journey continued. The river became narrow and the current swift. Now Klatok directed Cha it zit to steer the canoe in zig-zag fashion across the river from bank to bank to avoid a direct confrontation with the current. At times she had him hug a bank where the river eddied gently, or where they could grab trailing branches and pull the craft upstream. Cha it zit wondered if the day would ever end. His muscles were cramping, and he ached intolerably, but he would have died still paddling rather than complain.

The canoes had become separated, some, depending on the strength of their occupants, had pulled ahead of others. When the river flowed openly, they were visible to each other; at other times it was as if the canoeists were each alone on the treacherous river.

At a place where the current slackened and the water was still, Klatok eased her demands and the canoe drifted. "Listen," she whispered hoarsely. All the world seemed still, there was scarcely a sound. Then Cha it zit heard a thump, thump, thump. He looked at Klatok, a question in his eyes.

"Beaver," she said. "They are building their dams to provide lodges. The sound is the slap of their tails pounding mud and sticks together. Look," she pointed to where alder trees had been neatly cut off. Each trunk now a projecting point. "They have bitten off the trees and stored them in their winter lodges. Just as we do, the beaver people store food for the winter. We are all brothers."

Cha it zit had eaten soup made of fat beaver tails and he knew that beaver fur made the warmest robes, but he had not admired them until now. "I am proud to be brother to the beaver," he said.

The water was shallow and filled with trout. Looking down into it, the paddlers could see the beautiful fish floating lazily just atop the stones. Occasionally one would flash to the surface to snatch a fly that dipped too low.

"Trout would be good for dinner," Cha it zit said, wishing he had a dip net. Just then a plump rainbow trout leaped from the water and landed in the canoe. Cha it zit laughed and rapped it sharply with his paddle.

"I heard that you have a fishing spirit," Klatok said, "I witness that is true."

Cha it zit admitted to himself that it must be. Another canoe came in sight and Klatok held up the Trout. "See what gave itself to us," she called. Cha it zit grinned with pleasure. He was discovering that Klatok, the woman that everyone feared and respected – the 'strong one' they called her – was an enjoyable companion.

They continued up the river, sometimes paddling, sometimes poling. Without warning they rounded a bend to find that their way was barred by another jam of matted trees and forest debris. Cha it zit groaned in dismay. The lead canoes were already beached and being unloaded for the second portage. On the bank above the river stood a longhouse.

"Lokanum's house," Klatok said. "He is a rising young leader of the Nooksack people. Sax ump ki has signaled to him by now. They will let us pass."

The family of Lokanum, old and young, stood by their house watching the canoes come in and land. The Lummis unloaded quickly and carried their goods and canoes past the jam. There was no sociability, but neither was there hostility.

"They are different," Klatok said. "We cannot understand each other, but we are not enemies either."

The river became swift and narrow. A sound like muted thunder became audible. It grew in volume until the noise was fearsome. Next they came face to face with an awesome sight. The river had formed a great bend, almost a circle, which the water was gushing through with tremendous force.

"Big Nose," Klatok said. "Here we are very careful. Do as I tell you." Cha it zit noticed that the boiling, seething water was choked by many fallen trees. The water cascaded over, under, around them with angry force. The air was filled with spray.

Klatok sat firmly, using her paddle to indicate the direction Cha it zit must steer and to stabilize the plummeting, yawing craft. Often she chose to hug a bank, keeping away from the main thrust of the river. At times when the logs blocked her way, she directed the craft directly into the rushing current, sending it catapulting upstream and catching a cross current to maintain direction. Cha it zit was amazed at her skill. "This is some woman," he said to himself, truly appreciating his father's main wife for the first time.

When they were finally free of the clutch of Big Nose, they found themselves in calm water. The area had opened and they could see the mountains rising above them. Kulshan, now nearer, rose majestic in a robe of white to the east where the sun rises. A little farther along, they came to a cluster of houses sitting back from the river.

"The river wanders as it wishes," Klatok said. "The people live far back so that it does not enter their houses."

Somewhat farther on, they saw that the lead canoes had been beached and the people were setting up a camp. "At last," Cha it zit thought. His hands were battered; blood had dried on the handles of his paddle. He doubted that he could straighten his legs after hours of kneeling.

Klatok showed no weariness. Her stout, chunky body served her well. Once ashore she assumed control of the women's work. She produced a large horse clamshell from the bundles she had stowed in the canoe. It had been tied shut with nettle twine. Carefully she opened it and blew on the contents. There was a whiff of smoke. She placed a few bits of dead cedar twigs in it and blew again. Cha it zit could see glowing coals send out fingers of flame to consume the cedar. Then Klatok gathered more of the cedar branches and laid a fire on the beach. She slipped the coals, now burning brightly, onto the kindling. Soon a cooking fire was ready.

The sun had completed its circuit across the late autumn sky and was sinking toward the hills to the west. A wave of color danced across the sky; gold, rose, and lavender created a visual poem. For a short time the people stood and looked. Nothing was said. It was a moment to be enjoyed within one's own self. Then it was over and the sky turned purple-gray.

The fish in the river became more active. They leaped into the air and dropped back with a splash. A kingfisher perched on an overhanging branch watching. Then suddenly, with folded wings, he dove neatly into the water. A moment later he emerged with a small silver fish in his bill.

"Kingfisher is teaching us a lesson," Sax ump ki said as he reached for his bone hooks and stout nettle line. Soon a row of fish, skewered on spits, were roasting before the fire.

The song of the river rolling toward the sea brought pleasure to Cha it zit as he lay on its bank that night. His bed was moss, his cover was birch bark which he had peeled in strips from a fallen tree. The fire was dying, slowly reducing itself to a red glow, casting a rosy light on the people stretched out on the sand around it. An owl hooted from somewhere in the dark forest that encroached upon the river. In the far distance a wolf howled and was answered by others of his pack.

As soon as it was light, even before the red orb of the sun appeared over the white mountain, Kulshan, the people were up and preparing for the day's journey.

As they progressed upstream the river changed capriciously; there were rapids to contend with interspersed by placid stretches. In places it spread out like a lake and there were islands rising from the almost still water. At last they came to a place where a wide, shallow river flowed into the Nooksack. On high ground above where the water of the two rivers mingled, stood a large village.

"Nexwisjam," Klatok called to Cha it zit. "That is the main village of the Nooksack people. Their great leader lives there. He is very old, but his son, Humptalips, is ready to take his place."

The houses were different from the Lummi houses, Cha it zit observed. They were very low, so low that a man would have to stoop. He was about to ask about that when Klatok said, "The people dig down when they build a house. Part of it is below the ground. It gets very cold here in the winter. That kind of a house is easier to keep warm."

"Are we going to stop here?" Cha it zit asked. It appeared that Sax ump ki, in the lead canoe was preparing to land.

"Courtesy requires that we ask to go up the river. The old headman controls the river and all the villages on it," Klatok explained.

The stop was brief. Only Sax ump ki left the canoe. Cha it zit could see the old man, his hair white, a whisper of a white beard on his chin, gesturing as he spoke in sign language to Sax ump ki.

The flotilla swung north, going up the cold, silt-gray river. It soon became treacherous, sand bars appearing, some just below the surface, to block progress. Then the land began to rise toward the mountains, and the river rose with it, tumbling from a series of cataracts. Before long they came to an open meadow. The waist-high grass was lush and green and the trees were beginning to show gold and red.

"Here one may find elk," Klatok said, scanning the open ground. "We will wait till later, when we return, to hunt."

They were in the mountains. The river had split again and another large stream flowed into it. Above the confluence, the Nooksack dwindled. It became a small, friendly river flowing in a bed of gravel. The great bulk of the white mountain, buttressed by other peaks, loomed above them. Cha it zit felt small and an uncomfortable uneasiness overtook him. The air was different. He breathed deeply, but it didn't satisfy his lungs. He felt suddenly a yearning for the open beaches, the surge of the sea, the familiarity of his village. There were spirits, evil spirits, in the dark forests that blanketed the hills and rose halfway up the overpowering peaks. He knew that evil beings were everywhere in this beautiful, forbidding country.

They had reached a place from where they could continue no further. From a cliff far above, a curtain of water fell to the stony ground and became the river. The people pulled their canoes up on the water-smoothed rocks and carried their equipment and supplies to cache them in trees away from animals.

It was a pleasant place. There were bushes loaded with blue huckleberries growing at the river's edge.

"We will camp here," Sax ump ki said, grounding his canoe. "Some will climb the mountain for mountain goats. Some of us will hunt elk and deer. The women can pick berries. We can get bear too; they, too, will want the berries that grow here."

That evening after the women had set up their temporary shelters of cattail mats and cleared away the remains

of the late meal, the group held a council. Beside the gently murmuring river, by the glow of the dying fire, they made their plans. As was the accepted custom, Sax ump ki, as headman, moderated the discussion. Those who had the proper skills and power were to leave early the next morning to climb a high peak, one visible above the others, where mountain goats were known to feed. Their's would be a difficult task. The goats were wily and had the ability to climb rock faces where a man could not go. Once a goat was killed, the precious hide and horns, as well as the meat, must be packed down the rugged, almost vertical, terrain. Only the most daring hunter went after mountain goats. Sax ump ki and Cha wentz were among those who felt they were strong enough for the challenge.

Other game, bear, elk, and deer were more valued for food and could be found in the open valley of the river. What la cum, who was a fisherman at heart, agreed to participate in that hunt. During the conference, Cha it zit said nothing. He kept his own counsel. If the others wondered why he hadn't offered to go on a hunt, they, too, kept their counsel. The women listened, but were not included. They already knew that they would spend their time picking huckleberries, gathering mountain grasses, and drying strips of meat into jerky.

When the talk had lapsed into silence and the sky had become black, the people slept. Cha it zit alone sat by the river listening to its forever song. It seemed to be talking to him in a language he could not quite understand. It spoke of the ongoingness of life and of death, of the eternity of the world-forces that his people called spirits. The great rock faces of the mountains that towered above him, invisible in the blackness, spoke of timelessness too.

He was surprised at the drift of this mind, but then he considered his surroundings. "How could I not think such thoughts in surroundings like this?"

According to the elders who knew of such things, many powerful spirit forces dwelt in the mountains. The great Kulshan itself, was a spirit. He thought of the stories he had heard as a child of how Kulshan had three wives. Two of them stood beside him. He had seen them on the trip up the river. The third wife was arrogant and demanding. Finally, Kulshan became disgusted with her and sent her away. She became so angry that she rushed headlong, scooping out the valley that the river ran through and hollowing out the land, tossing it up in heaps—creating the inland sea and islands that the Lummi people occupied.

"Here I am," he reflected, "sitting at Kulshan's feet."

Then he recalled another story. The storyteller had said, "This happened when my father was a boy. It is true." Cha it zit mulled over the details that his mind dredged up, half-forgotten bits and pieces. The people were camped on an island from which they could see the white cone of a mountain. Then the mountain had no name. Suddenly it exploded, fire shot from its top and ran down the sides. The smoke and ashes filled the air. The sky became so dark that the people could not see. Then lightening struck the mountain and thunder reverberated among the hills. The darkness lasted for days. The people named the mountain Komo Kulshan, he recalled, which means fire on the mountain.

"Be calm, Komo Kulshan," he said aloud. Only the river heard, as it continued its incessant song on its eternal route to the sea.

The hunters left in the morning, some to climb for mountain goats, some to range along the river for other game. The women gathered their berry baskets and wooden combs for removing the berries from the branches. No one said, "What will you do?" to Cha it zit. It was assumed that each would work according to his or her skill.

Cha it zit watched them go. Sax ump ki had left a hunting bow with a quiver of arrows in his dugout. On impulse,

Cha it zit borrowed them. His own knife was hanging in a sheath from a belt around his waist. He slung the quiver over his shoulder and grasped the bow. He wasn't quite sure why he did so, since he had no intention of hunting. Passing What la cum's canoe, he saw his uncle's fishing line and hooks. "There are trout in the river," he thought. "I'll catch some for tonight's meal."

He left the campsite, exploring his way upstream toward the falls. They fell in one clean drop from a sheer cliff so high that Cha it zit had to bend backwards to see the top. At the foot of the falls was a deep pool, worn down by countless years of countless drops of water.

He laid down the hunting bow and looked for bait. Any flying insect would do. A black fly, the kind that pester deer and elk until they nearly go mad, landed on his hand. He slapped it and baited the hook. Before long he had a pile of mature rainbow trout lying on a mossy log in the shade. He marveled at their beauty as they came from the water, silver, shining in the sun, with glints of green, blue, rose, and violet ranged rainbow-like along their sides. As they died, the rainbow faded and life drained from the silver.

Suddenly he heard a sound behind him, a snort unlike any he had heard before. Whirling, he saw a giant grizzly standing erect like a man, scenting the air. Their eyes met, grizzly and man. Cha it zit did not run; there was no point in it – the bear could outdistance him in minutes. There were no trees to climb and that, too, would be senseless, since the bear could out climb him.

In a moment, he knew, the bear would charge. The bow lay between him and the bear – could he get it? Cha it zit felt a tug on the line he was holding. He jerked it hard. The fish flew up and off the hook, landing close to the bear. For a second, the huge beast forgot him and lunged for the trout, still flopping on the ground. That was all Cha it zit needed. He ran for the bow, grabbing it, and came almost face to face with the grizzly. The bear slapped with his razor sharp claws. They grazed the boy's face. Cha it zit fitted an arrow to the bow and shot straight into the bear's chest. With a roar of rage and pain, the grizzly went backwards and lay bellowing and thrashing on the ground.

Cha it zit was dismayed. His arrow was embedded in the chest of a brother creature. "Bears are brother to man," the elders said. "When they take their bear coats off in their spirit existence, they are like us." He watched the death agony of the bear, and sat with him after he was still. "Go to your spirit world," he told him. "Do not hold what I did against me."

Then, a long time later, he began to skin the grizzly. It was a young male in a prime coat. The fur was glossy from eating fish and was a soft gray tipped with silver, the finest pelt Cha it zit had seen.

The skinned body of the bear did look human but Cha it zit forced such thoughts out of his consciousness. He cut the meat from the bones and then into pieces. These he wrapped in the skin, leaving the bones for the carrion eaters.

"Tonight," he knew, "the coyotes and wolves will come. Tomorrow the ravens and eagles will have a turn."

He brought a canoe to the pool, loaded the meat into it, and let it drift on the current back to the campsite. After that, he set out to find the elk hunters, but there was no sight or sound of them.

It was late afternoon when he returned to the camp. He could hear the happy chattering of the women now bringing in baskets of the sweet blueberries. They shouted to each other in amazement when they saw the grizzly pelt filled with meat. "Where did it come from?"

"What a beautiful fur."

"I shot it," Cha it zit said, trying to sound manly, casual.

"You?" they questioned, eyes wide.

"You said you would never be a hunter," Klatok's voice was almost accusing.

"People can change," Xedewa said, coming to his aid.

"This one certainly did," Klatok agreed reluctantly.

The hunters returned with several young deer and beaver, which they sat about skinning. "We found no trace of elk," they said. "They haven't come down from the high country yet. We need snow to bring them down."

"Do not wish for snow until my husband and son are back," Klatok said in a harsh tone.

Those who went for mountain goats often stayed for a few days. The people in the camp watched the sky and tested the wind. A sudden storm could bring torrents of rain, or snow, with temperature drops so drastic that a person could freeze in a short time.

Each day the hunters went out and each night they brought back several deer and an assortment of smaller animals. The women now spent their time cutting and drying the meat over low, smoky fires. There had been no sign of the goat hunters. The people showed tension. No one expressed concern, but it was present. It could be seen in their eyes, in their silence. At last Cha it zit could stand it no longer. "I am going to track them," he announced.

"I'll go with you," What la cum offered.

The next morning, as soon as the first glow of the sun appeared behind the mountains, the two scouts left the camp. They wore little clothing and carried only their knives, so as to be burdened as little as possible for the arduous climb ahead.

The trail left by the hunters through underbrush in the lower elevations was difficult to follow, being many days old. Here and there the two searchers found bent or broken branches and in soft places, where ground water seeped close to the surface, they found footprints. They reached the base of the mountain peak that Sax ump ki had said they would climb. Here the hunters had found a deer trail and the tracking was easy. Cha it zit and What la cum then climbed at a steep angle through huckleberry bushes drooping with ripe fruit. Gradually the vegetation changed. The bushes became dwarfed, misshapen, hugging the ground. Mountain blueberries replaced the huckleberries, growing in low masses, offering their sweet berries to bears and birds. Higher up, the stunted mountain ash had changed to its fall colors of red and gold.

After hours of steady climbing, the pair was in open country. No trees grew at so high an elevation. They stopped to breathe in the thin mountain air, filling their lungs, but not satisfying the needs of their bodies. Above them a marmot sat on a rock and whistled an alarm. "They are curious about us," What la cum said.

"Afraid, too," Cha it zit replied. "If we could talk their language they might tell us where the goat hunters are."

Below them they could see the river and the campsite. The people looked like bugs as they moved about their tasks. Beyond and yet beyond, mountains rose, row upon row, as far as their eyes could see.

"So much space," Cha it zit said. "I didn't know there was so much space. What is in it all?"

"Some say that strange beings inhabit that land. They are to be feared"

"Spirit beings?"

"Maybe. But others, too." What la cum looked directly at Cha it zit. "Like Sasquatch."

"Sasquatch?"

What la cum nodded. "I know those who have seen them. One of the Sasquatches carried a relative of mine, a woman, to his home and kept her there for many days."

"What is a Sasquatch like?"

"Bigger than a man, and stronger. Covered with fur, which is silver-tipped, like your grizzly. They walk like we do. If you put on a bear coat you would look like a Sasquatch, only not so big."

Cha it zit thought for a moment. "I heard of such creatures as a child and I was frightened then. Now I am a man and should put such fears behind me," he said, trying to subdue the uneasy feeling that had come over him.

"I have never heard of one harming anyone," What la cum commented. "They want to be left alone. No, I do not fear them, but I don't want to meet a Sasquatch, either."

"Then I would like to see him, this Sasquatch."

"Cha it zit, I believe you have strong powers, so do not wish to see a Sasquatch," What la cum said.

They climbed farther. Above them was a rocky pinnacle, the top of the mountain. Still there was no sign of the hunters.

Cha it zit called out, "Sax ump ki." The sound of his voice echoed from mountain to mountain and came back to him. After the sound faded, from somewhere in the distance, he heard an answering shout.

"Come!" he said to What la cum as he dashed toward the voice.

They followed the calls, shouting and listening, across a narrow ravine to the other side of the peak. There they found the hunters huddled together around a small fire. Several mountain goat carcasses lay a short distance away.

When he was closer, Cha it zit noticed that the men seemed unnaturally quiet. He saw ugly bruises and cuts on their bodies. They made no effort to rise, and, while they greeted Cha it zit and What la cum, their voices were dull and expressionless.

"They seem to be in a trance," What la cum said, his face puzzled.

"Or in great pain," Cha it zit added.

"Could they have seen a vision?"

"Let's ask them," Cha it zit said, ending the speculation.

“What happened?” What la cum spoke directly to his brother in a loud voice, pronouncing each word distinctly.

Sax ump ki shook his head in confusion. He groaned. “I’m not sure. None of us are, but I’ll tell you what I seem to remember.”

The others nodded but made no effort to enter the discussion.

“We had come around the peak when we sighted a small herd of goats. They were beyond that slide.” He pointed to an escarpment that jutted beyond a place where rocks had loosened and slid down a steep bank. “They bolted when they saw us but we managed to hit several of the big horned males. Then we went to get them and carry them out. We were careless. I suppose it was the excitement of killing them. The rocks were lose and we started a slide. Down we went, the six of us and the dead goats. That is the last that I can remember. I must have been knocked unconscious by the rocks.” He looked at the others. “Can any of you tell what happened next?”

Cha wentz, who had been lying flat, his leg at an odd angle, pulled himself up painfully until he was resting on his elbows. “I recall something very strange,” he said. “I think I was awake, but I can’t be sure. Everything was fuzzy and seemed to be going around slowly. I hurt. I tried to move but I couldn’t. There was a weight on my leg. Then I felt the weight move and it was gone. I felt someone near me. Then I was lifted from the rocks and carried. I tried to see who it was but there was fog in my eyes. It felt like an animal, covered with fur, but it a carried me as a man would. It was gentle and put me down here. I cannot move. I think my leg is broken.” Then he opened his fist and showed long silver hair lying there. “I must have pulled it out when I was being carried.”

Sax ump ki interrupted. “It even brought up the goats.” He pointed to the carcasses piled one on top of the other.

"Sasquatch?" Cha it zit asked.

"Who can say?" Sax ump ki wondered.

"What else could it have been?" What la cum asked. "What man could be that powerful?"

The injured men shook their heads.

"We have to get you down," Cha it zit said, changing the subject to a more practical one. "But how?"

"Cha wentz will need help," What la cum said. "How about it? Can any of you walk?"

"If you can make us walking sticks, we can try," Sax ump ki suggested hopefully.

"You are younger," What la cum said to Cha it zit. "Find some stout sticks down on the lower slopes. I'll see if I can do something about Cha wentz's leg."

It was several hours before the hunters were provided with sticks and Cha wentz's leg was splinted. Then with torn muscles, sprains, and bruises, the men began the painful descent.

Progress was slow. Cha wentz leaned heavily on Cha it zit's shoulder as he hopped along on his one good leg. Others, assisted by sticks used as crutches or canes, gradually worked their way down the rough, steep deer trail. What la cum, at the rear, assisted when he could and urged the hunters on. It was nearly dark when the pathetic party reached the camp.

When she saw them coming, Xedewa set out to find healing herbs. Klatok stirred the fire and set cooking rocks to heat. Others ran to the river for fresh water to wash the men's wounds or brew a bracing tea.

"How did this happen?" the women asked. Again the story was told, this time with embellishments which added excitement to the tale and glory to the hunters. The Sasquatch became an active participant in event. When the women expressed wonder, the men swore it all happened as it was told.

The next morning, Cha it zit and What la cum, accompanied by Klatok, who insisted on seeing the spot where the misadventure had occurred, again climbed the mountain to retrieve the slain goats.

They found them as they had left them but Cha it zit noticed that the area had been disturbed. Branches that had been torn from the saplings that had been used as canes had been carried away. The area around the fire pit had been swept clean. Except for the goats, it looked as if no one had ever been there.

"If I didn't believe that a Sasquatch had been here, I would now," Cha it zit said.

"No bear did this," What la cum agreed. They gutted the goats and pulled the skin down from their bodies. Klatok was preparing to cut the meat from the bones when What la cum stopped her.

"We cannot carry it all down the mountain. Let's take the hides and the horns but leave the meat. Someone or something may come and take it."

"Yes, let us potlatch the Sasquatch," Klatok said, her eyes twinkling.

Even carrying only the hides and horns down the steep terrain was difficult. The skins, oozing juices, constantly slipped from their shoulders and the heavy, bulky horns were hard to grasp. The three lost their balance often and slid or rolled down the slope. By the time they arrived at the river, they appeared as battered as the others were.

It would be several days before the injured hunters were able to travel. Even then it would be a difficult trip, for the river was more treacherous going downstream than it had been when ascending and every man was needed to guide the now heavily loaded, always clumsy, river canoes. But with six of the strongest injured, the prospect was grim.

The women nursed and fussed. Xedewa, particularly, hovered over the patients, dosing them with concoctions which she assured them would ease their pain and heal their bodies. The last of the meat strips hung drying. Already most of the stores had been tightly packed, ready to be transported. Berry cakes, huckleberries mixed with rendered bear fat and dried hard, were in baskets; and jerky was in watertight boxes, with scraped hides in bundles.

Sax ump ki watched the sky. On the third day after the rescue, a few flakes of snow fell. The mountains that had seemed so inviting, offering their richness, now looked cold and forbidding. He made a decision. "We cannot wait longer. The wind of the north is about to take control. We must leave."

The people were relieved. It was foolish to wait longer and they knew it. "Let those who are injured take places in the bow," he said. "Klatok and Xedewa, you are strong – let us see how strong, take a man's place in the stern. You too, Cha it zit. You are inexperienced, but you will learn."

The people pushed their canoes into the current and moved rapidly downstream. There were harrowing places where a canoe seemed to completely disappear in a swirl of water, only to rise again on the crest of a wave. Sax ump ki led, calling encouragement to those following.

They swept along, the river providing momentum, past the village of Humptalips', past rapids, through Big Nose, and then portaged the little and big jams. It was exhilarating to Cha it zit. Each moment presented a new challenge. His every sense was alert, his body responding to their silent, urgent messages.

Incredibly, they reached the place where the river split with no accidents. Now they were in their own territory drifting easily along on their own river. Cha it zit realized that he was very, very tired. He relaxed his grip on his paddle and stretched his fingers. "It is good to be home," he said aloud.

14

The Cycle

Sta nek was sitting beside the cooking fire when Cha it zit entered carrying the grizzly bear pelt. The old man seemed to be drowsing, but he broke into a toothless grin when Cha it zit called to him.

"Ha!" he cackled. "The hunters return. Did it go well for you?"

"Well enough," Cha it zit answered, dropping the heavy skin on the platform. "You will hear about it. There is much to tell."

Wanana moved toward him, pleasure in her eyes. "It is good to see you, my son," she said softly.

"We have brought much food," he told her, although his smile said much more.

"A grizzly pelt," she exclaimed, examining it with admiration.

"I shot it."

"You? My son who would not be a hunter?"

"Yes."

He did not tell her more. There was time. In the winter, when stories were told, then he would tell his.

The injured hunters, using their sticks, limped into the house and became the center of attention. The women clucked over their wounds and the children stared.

"They will have stories to tell, too," Cha it zit thought as he picked up the grizzly fur and carried it to his own place.

A mat had been hung, separating it completely from the rest of the house. He pushed it aside and there sat Tsexad. He noticed that she was working on the basket she had flung aside when she ran from him. It seemed so long ago.

He did not speak. He only looked at her. "How beautiful she is," he thought. "But how thin and pale." There were dark shadows under her eyes, eyes which seemed to have grown larger in her pinched face. Her arms, once round and strong, were like bones that had washed up on the beach. She met his look directly, not lowering her eyes or bowing her head as she usually did.

"What happened to you?" he asked when he had recovered his composure.

"Let Tseutsi tell you," she answered.

"You are thin. Have you been ill?"

"I have been dead and am now alive," she answered. "What of yourself?"

"There is much to tell," he answered, not sure how to speak to this woman who looked like Tsexad, but who spoke like someone else. "There will be time for that later."

She touched the grizzly pelt. "It is green," she said, "I will tan it for you. Did you kill it?"

"Yes," he answered, but offered no details.

"Are you glad to see me?" she asked.

"Yes," he answered. He could think of nothing that seemed right to say. He was confused. He had to think. "I have things to do," he said as he left.

Sta nek was sitting exactly as he had been when Cha it zit left him. "How old he has become," he thought to himself. "Where is Tseutsi?" he asked aloud.

"Somewhere with Latsi," Sta nek answered in a petulant tone. "She spends all her time with Latsi. I never see her anymore. That old woman has forgotten that she has a husband."

"Latsi. Of course! She would be out of her confinement and needing care to regain her strength. I wonder if she had a vision?" He thought of his own spirit quest, of the time of training and then the actual experience. "I would rather go through that again than endure what she had to," he decided.

"Where is Latsi?" he asked his mother. She had taken the baby from his cradle board and was gently massaging his small body. The marks of the head-flattening stone were on his forehead and his eyes bulged like a frog's, but he was strong and he held a small hand out to Cha it zit. Cha it zit touched it with a forefinger and the baby grabbed the finger, holding tight.

Cha it zit laughed, "Hey, you are a strong one," he said. "Someday I'll teach you to fish and gamble."

"He likes you," Wanana remarked. "Latsi has her own place now," she said. "It is beside mine." She pointed toward a partition of mats directly across the house from his own.

He stood outside her cattail mat divider and called, "Latsi."

For a moment there was no answer and then Tseutsi asked in her used-up voice, "Who wants her?"

"Cha it zit."

"You have returned. She has been asking for you. It is unusual, but come in."

Latsi lay on a bed of furs. If Cha it zit had not known it was her, he would have thought he was looking at a ghost. She was thin beyond life. Her body was wasted to nearly nothing. Her lips were blue, her hair hung limp and lifeless. She stirred when she saw him. She regarded him with large, dead eyes.

"Latsi!" He was dismayed. Then he turned to Tseutsi and demanded, "What did you do to her?"

"Do?" Tseutsi shouted at him. "Do?" I helped her have a vision, a great vision, one that will bring her prestige and power."

"Will that help if she dies?" He was furious.

"You went through it. Why are you so angry?"

"She is a girl. She is not strong. Look at her."

Latsi tried to sit up, but fell back. She raised her hand to stop the tirade. "Don't be mad," she spoke in a voice so weak that Cha it zit had to bend over to hear her. "She tried to help."

"She can't eat," Tseutsi complained. "She is all dried up inside."

"I'll get Xedewa," Cha it zit announced. "She will know what to do." Then he thought of Tsexad. "Grandmother, what of Tsexad?"

"Do you know what happened?" Tseutsi countered defensively.

"No. I'm asking you."

"When Latsi was placed in confinement Tsexad came begging to be kept there, too. She was afraid of Cha wentz and she was upset with you. By the time I found her there with Latsi, she was already contaminated by the spirit power in Latsi so I couldn't tell her to go away. Where could she go?"

"She went through the same ordeal that Latsi did?"

"Yes."

"Did she have a vision too?"

"A very strong vision. It was so powerful I thought it was going to be too much for her woman's body. She threw herself around so much I had to sit on her."

"She is different," he shook his head.

"Yes," Tseutsi admitted. "She has strong powers."

Cha it zit found Xedewa, who hurried away to find medicine to open Latsi's stomach so she could eat. "It will take a tea of boiled fir bark," she said as she ran toward the forest.

Cha it zit returned to his place. Tsexad was still occupied with the basket. She looked up as he entered.

"I saw Tseutsi," he told her. "Now I know what happened to you. Did you realize that I would look for you when you hid?"

"I hoped you would," she said candidly. "I wanted you to worry. I wanted to hurt you like you hurt me."

"Then you weren't hiding from Cha wentz as you told Tseutsi, you were hiding from me?"

"Yes."

"Why?"

"Don't you know?" she became upset, she raised her voice to him, something that a well-mannered woman would never do. He reacted, anger rising in him; then he controlled himself.

"Why don't you tell me?" he suggested.

"Because you are to have a wife. I told you how I felt."

He put his hand on her shoulder and then he comforted her as one would a hurt child. She held on to him.

"I will never let you go," she said.

"Nor I you. Tsexad, you are a woman now. You have spirit power of your own. In my eyes you are not a slave. If you belong to me it is because you want to. If I belong to you, it is because I need you. This is my promise to you, no other woman will take your place. It will always be reserved for you."

She smiled, "That is all I ask."

As he often did, Cha it zit sought his place of refuge, the rocky ledge above the sea. His thoughts tumbled over each other, coming at random from the inner recesses of his mind. He tried to sort them out. So much had happened to him he hardly knew where to start. One thing he did know – he was supremely happy.

The spirits had been good to him. He had proven that he had a strong gambling spirit to help him become wealthy. Somehow, he had also attracted a fishing spirit gift. He was now a man among men, looked up to, respected. One day he would have a wife from a strong family and a house of his own in a spot he chose, at Temxwigqsan, near the place of his power vision. He knew, too, that if he wished he could assume the shaman powers that Tselique had passed on to him.

But another thought came to him like a dark cloud blotting the sun. The white aliens. What of them? Would they go back to wherever they had come from, or would more and more arrive bringing their boxes of strong water and wanting furs? What of their *thunder sticks* that kill?"

"I fear for the future," he admitted. "I see evil days ahead for us. How can we stop something that we don't understand? And yet, maybe the strange beings have good things, strong powers, to give us."

He recalled how Xedewa had said, "To fight them we must get their powers." Tsexad had spoken of the 'God' of the beings who made salt at a great river to the south. Some of the people were even now searching for that 'God' far to the east where those aliens, Lewis and Clark, had come from.

He thought of his promise to Tsexad. Could he keep it? She was satisfied. That was good, but what of his duty to the family, to the wife he must take?

His mind went round and round, but he found no answers, only questions. There was a shuffling noise behind him. He turned to see Sta nek standing there, holding a branch for support. His hands shook and he looked tired.

"Sit here," Cha it zit offered.

"Grandson," the old man said when he had gotten his breath, "I will not be with you much longer. There are words I would leave with you. Listen, for this is important to me. It may be important to you too."

"Speak, grandfather."

"I have watched you grow from a boy to a man. I gave a potlatch for you when you received your name. Now I see you as a true man with spirit powers that will bring you wealth and respect. That is good.

"I am old, I have dreams. In my dreams I see that our lives, lives that we have led from the earliest days, are changing. The old ways will pass away. I see that in my dreams.

"I see, too, that the leadership of our people will pass to you someday. You must be wise and you must be strong if you are to help our clan fit into the new way of things. I will not be there, but I know these things. I have seen them as a vision." Sta nek stopped talking and seemed to lose himself in his thoughts. Cha it zit waited quietly, thinking that when he was ready the old man would continue speaking what he had come to say.

Sta nek coughed and cleared his throat. "The Yakultas will grow bold. They will come to our shores with the *thunder sticks* we have heard about. They will grow in power. We must be able to rid ourselves of those marauders. Yes, we must get *thunder sticks* too."

"How can we do that?" Cha it zit wanted to know.

"Only the aliens can give us such weapons. We must find those strange beings and we must give them what they ask for so they will give us *thunder sticks* in return."

"They seem to only want furs," Cha it zit observed.

"They have other gifts, too, good gifts and bad gifts," the old man continued. "You must be wise and choose the right ones. Much will depend upon you. I fear the aliens, as I fear the Yakultas, but we need the aliens' power. I fear that the aliens may yet be our enemies. I fear that they will come and come, as the waves of the ocean. They will flood our lands, there will be no place where they are not. There will be no place for us in all this vast land and wide sea. Where will the spirits be? Where will our power lie?" His old voice quavered and was still.

Cha it zit put his hand on Sta nek's shoulder. Together they looked out toward the islands, now shrouded in the darkening night. "I have been thinking the same thoughts," he said, his voice soft and low.

"I tell you these things so that you will prepare yourself for the days ahead. I will not be here, but my words will stay with you."

"I know that what you say is true," Cha it zit responded. "I don't know what is going to happen, but I fear for myself and our people. I promise you that if leadership falls on me, I will remember your words. I will remember what you and another, whose name I cannot mention, have taught me. But go in peace, grandfather, the problem will not be yours."

"Yes. Yes, I know. I worry for you, my grandson. I worry for you."

They sat there together. The old man whose life was behind him and the young one whose days were only beginning. The sky darkened, but still they sat, each reluctant to let go of the communion he felt with the other.

"The night is black," Cha it zit said at last. "Will it be black for our people in the time that is to come?"

"I do not know, but there is more that I wish to tell you," Sta nek spoke into the dark silence.

"Yes," Cha it zit answered softly, "I would hear your words. There is much that you can teach me." Then he added reluctantly, "the time left for you here may be short."

"Already I can feel the breath of the spirits who will take me. But I am ready. My life has been long and I feel very tired. I have learned many things. Now I will give what I know to be true to you."

"Speak, Grandfather, I am listening."

"Since the time, long before the earliest memory of our ancestors, when the creator made us and all the world around us, and his son *the Changer,* came to put all living things into their present form, our people have perfected a way of life. It is good. Even today we repeat the same customs and the same ways of our forebears. That is our heritage. It must not change lest we cease to be *the People.*

"The old ones knew the time when the ducks would come and the herring and the salmon. That knowledge is precious to us. They learned how to preserve the foods that they gathered so that even today we do it the same way and we are not hungry. The women make baskets as they learned from their mothers who learned from their mothers before them. They must not forget to teach their daughters the art, or it will be lost. So it is with all knowledge. We receive it, and later we give it."

"It will not change," Cha it zit assured Sta nek.

"My visions – my dreams – tell me that all will change. In a few circles of the seasons, the knowledge of many generations will be gone. Then our people will be no more."

"No!" Cha it zit answered sharply. "Your dreams are wrong. That cannot happen. Our ways are full of power. We are strong!"

"Save your anger for when you will need it," Sta nek said gently, placing a bony hand on Cha it zit's shoulder. "The circle of my life is about to close, but I will continue in the same fashion. As a circle never ends, I will always be. Each year the spirits come in the winter, and then leave to circle the world until the next winter. Salmon come at their usual time and then are gone, but they always return. It is so with the geese and the ducks, and it is so with our people.

"We quest for food, as all the generations before us have done. Each part of the year's cycle finds us in a certain place. There we gather food that *the Changer* provided for us in ancient times. That food, too, has its own circle. All things live and die in their circles and all the circles mesh. Each depends on the others."

Sta nek slumped and his hand slid from Cha it zit's shoulder.

"Grandfather, are you all right?" Cha it zit asked in alarm.

"Don't let the circle break," the old man muttered as he slid from the rock and lay motionless on the ground.

Cha it zit picked the old body up. It seemed to be weightless it was so thin. He hurried toward the longhouse with his burden.

"A circle has closed," he said to the family of Sta nek as he laid the tired old body on a cattail mat. He left the grieving to the women. His own sorrow he would bear in private. He knew that he would long cherish and ponder the words that Sta nek had spoken to him.

A blanket of grief fell heavily upon Cha it zit. The atmosphere of the longhouse was oppressive to him. Tseut-si was rocking and moaning beside her dead husband, other women were wailing, and the children cowered in a corner.

Sax ump ki and What la cum were conferring quietly and Cha wentz paced between them and the women, not knowing what to do. Unobserved, Cha it zit slipped outside into the silence of the night.

"Don't let the circle break," Sta nek had told him. Cha it zit knew that the wise old man had felt a coming danger. No doubt it was a fear of the strange aliens, the same apprehension that he also felt. He felt proud and pleased that Sta nek had chosen to pass his wisdom on to him and not Sax ump ki or Cha wentz, even though they were more entitled to it.

"He saw me as a leader, worthy of his words," Cha it zit told himself. Then he realized that what Sta nek had believed was true. "I am a leader," he said to himself. "I will prove it. The circles of our lives will not be broken. They will not be broken by aliens or any other force. I promise you this Sta nek. For as long as I live, I promise you this."

His feelings became so intense that for a moment reality was blocked out. Then a vision came to him. He saw, out of the darkness of the night, a white fog approaching across the sea. As it came nearer, he heard strange sounds – voices speaking words he could not understand, boomings, songs, poundings. Then the fog was transformed into many faces, all aliens. On and on it came toward him, until he was engulfed in the smell and sight and sounds of countless pale-faced, whiskered aliens. He stood defiantly among them. He spoke to them, eloquently, begging for his life and for that of his people, but no one heard him. He felt stifled. There was no air for him to breathe. Then came oblivion.

Cha it zit told no one of his vision. But he felt its power. He knew that he had been transformed and never again would his life be as it had been. He was destined to be a leader, he knew for a certainty, and if he was to keep his promise to Sta nek, he would have to devote his life to his people.

Epilogue

The story of Cha it zit and his people has only just begun. There is much more to write of the life and times of this remarkable man, one of many wise and courageous Salish Indian leaders.

With the coming of the Hudson's Bay Company's fort and trading post to Langley, in what is now Canada, on the banks of the Fraser River, to the appearance of hoards of prospectors searching for ways to reach the goldfields of the north, to the presence of the missionaries, along with settlers and rum runners, the early and mid-nineteenth century became a time of traumatic and often devastating change for the Lummi people.

Leaders like Cha it zit stood tall as they relied on the wisdom of their elders and the values of their heritage to face the onslaught. They guided their people through a morass of governmental indifference, misunderstanding, and poor judgment.

It is a wonderful, exciting story, all the more wonderful as one leaves the field of conjecture and relies on historic events as they were recorded. It is my hope and my intention to continue to search for and write the life story of Cha it zit, a leader who bridged the gap between two cultures.

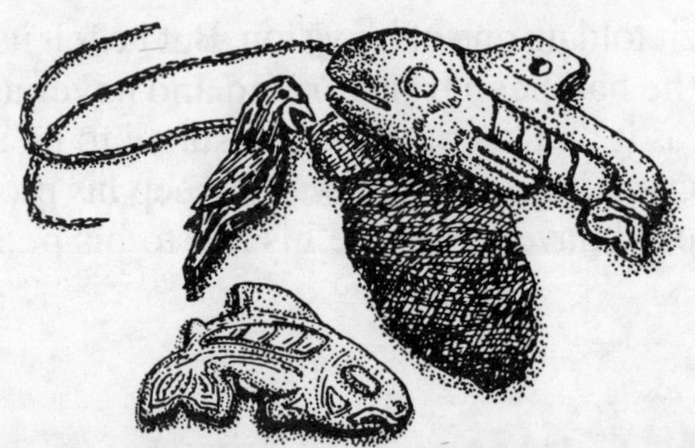